# Robotic Control

**P. M. Taylor**

*Department of Electronic Engineering*
*University of Hull*

Macmillan New Electronics
Introductions to Advanced Topics

MACMILLAN
EDUCATION

First published 1990

Published by
MACMILLAN EDUCATION LTD
Houndmills, Basingstoke, Hampshire RG21 2XS
and London
Companies and representatives
throughout the world

Printed in Hong Kong

British Library Cataloguing in Publication Data
Taylor, P. M.
Robotic control.
1. Robots
I. Title
629.8′92

ISBN 0–333–43821–3
ISBN 0–333–43822–1 pbk

# Contents

# Series Editor's Foreword

The rapid development of electronics and its engineering applications ensures that new topics are always competing for a place in university and polytechnic courses. But it is often difficult for lecturers to find suitable books for recommendation to students, particularly when a topic is covered by a short lecture module, or as an 'option'.

*Macmillan New Electronics* offers introductions to advanced topics. The level is generally that of second and subsequent years of undergraduate courses in electronic and electrical engineering, computer science and physics. Some of the authors will paint with a broad brush; others will concentrate on a narrower topic, and cover it in greater detail. But in all cases the titles in the Series will provide a sound basis for further reading of the specialist literature, and an up-to-date appreciation of practical applications and likely trends.

The level, scope and approach of the Series should also appeal to practising engineers and scientists encountering an area of electronics for the first time, or needing a rapid and authoritative update.

# Preface

Robots have already become standard tools for use in industrial automation systems, mainly where a variety of simple repetitive tasks is carried out. Their key feature, programmability, also allows their operation to be modified according to sensory information about their environment. Their surroundings may now be less well-defined, opening up new application areas and reducing the need for specially made jigs and fixtures. However, there is a long way to go before they can do useful tasks in completely unconstrained environments and before they have intelligence, instead of operating on explicit pre-programmed instructions. Once this has been achieved, the potential of robots will become enormous.

An initial study of robots would comprise an investigation into their mechanical construction, a survey of the types of sensors and actuators which might be used, and how control might be achieved so that the robot can perform the desired movements with the required speed and accuracy.

But practical robotics is much more than this. The robot must interface with its environment, in which there could be objects to be handled, feeding devices, other robots and people. A robot will usually be part of a larger system which can be programmed to perform a multitude of tasks. All of these items are considered here as being part of *Robotic Control*. The aim of this book is to present the concepts and technologies used in state-of-the-art robots, with an emphasis on fixed-base robots for industrial applications.

The early chapters describe the structure of a robot and its hardware components, followed by a study of how good dynamic control can be achieved. The programming aspects are considered in chapter 5.

The second part of the book develops the theme of adding external sensors and other equipment to the simple robot, making possible more advanced applications. Chapter 6 starts with an overview of typical external sensors and sensory data processing, and then looks at how this information may be used to enhance the capability of a robot. Workcell mechanisms such as feeders are described in chapter 7 which culminates in an illustrative example of the design of a sensory robotic workcell. Finally, chapter 8 attempts some crystal-ball gazing at future trends, based on reports of work being carried out at research institutes around the world.

The book is intended as a text for undergraduate courses or as an introduction for graduates new in the area. A basic working knowledge of computers and programming is assumed throughout, and of Laplace transforms in parts of chapters 3 and 4.

# Acknowledgements

The author wishes to thank his parents for their early encouragement to pursue a career in engineering. He also wishes to acknowledge the continual support of Gaynor, his wife and colleague. Professor Alan Pugh at Hull University must also be mentioned for his help at the start of the author's research career in robotics.

All photographs are courtesy of the University of Hull unless noted below.

Figures 2.9, 2.10 and 2.13b courtesy GEC–Marconi Research Laboratories and Alvey 'Design to Product' project.

Figure 6.14 courtesy of the Department of Computer Science, University College of Wales, Aberystwyth.

Figure 7.7 courtesy of SATRA Footwear Technology Centre.

# 1 Introduction

Many people are fascinated by the operation of mechanical devices, particularly those which partially mimic human behaviour. This fascination seems to have been prevalent throughout the ages. It was common practice, for example, to adorn mediaeval clock mechanisms with mechanical figures which would seemingly parade and strike the bells as the clock reached the hour. Even today such mechanisms attract many sightseers. However, they are essentially only simple sequencing machines: the same series of events will take place time after time in a pre-determined and unalterable order.

Robots have the same fascination but the control needed for robots is far more extensive than that needed for simple sequencing machines. Not only will a certain sequence have to be carried out, but its operation must be insensitive to external disturbances and variations in the dynamic characteristics of the mechanisms. In this way the task can be repeatedly performed with the same precision. The framework for achieving this aim is provided by the study of automatic control.

In addition, the sequence of operations must be reprogrammable so that different tasks may be undertaken. It is the need to perform reliably and cheaply a wide variety of tasks that underlies the use of robots. As the task capability is paramount, we will take the term *robotic control* to cover not only the control of the mechanisms of the robot but also the associated sensory systems and other mechanisms needed to carry out the task.

## 1.1 Historical development

The concept of feedback lies at the heart of control engineering. The onset of the Industrial Revolution provided a potentially powerful means of driving machines — steam. However, it was also potentially very dangerous, exploding boilers and engines being not uncommon, and there was a great need for some way of automatically regulating the steam supply according to the load.

Sir James Watt's flyball governor, introduced in 1787, provided one solution to part of this problem. As shown in figure 1.1, if the load on the output shaft suddenly decreases the shaft will accelerate, causing the centrifugal forces to move the rotating balls outwards. This then raises the

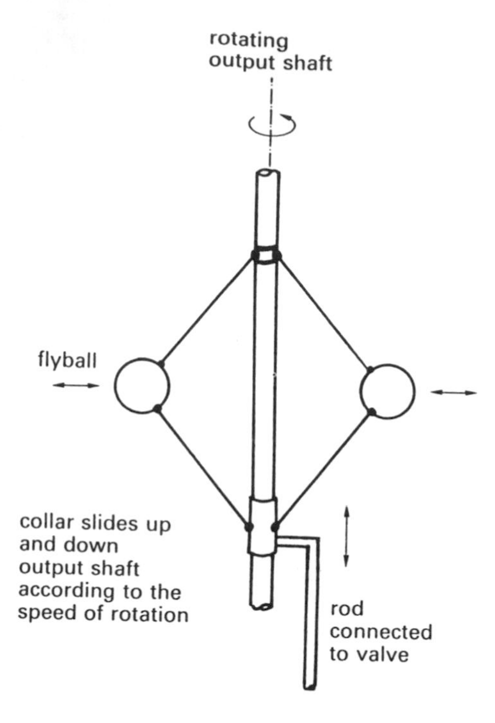


Figure 1.1 Watt's flyball governor

sliding member connected to the steam valve, thereby decreasing the steam supply and eventually reducing the output shaft speed. This principle was in fact in earlier use in windmills (Mayr, 1970) but the steam engine provided the first widespread application. A detailed treatment of the early history (up to 1930) of control engineering in general may be found in Benett (1979) and Mayr (1970), and of steam engines in particular in Dickinson (1963). As so often happens, it was only later that a theoretical analysis of the dynamics was attempted (Maxwell, 1868, reproduced in Bellman and Kalaba, 1964) once problems of unstable behaviour had arisen and required a solution.

The key development in the 1930s was the analysis (Nyquist, 1932) of stability in the new electronic amplifiers. An excellent collection of key papers on this and later developments in control theory has been assembled by MacFarlane (1979). This use of feedback was quickly applied to mechanical devices so that loads could be moved rapidly and accurately to pre-determined positions.

The industrial robot therefore embodies both the principle of sequencing and the use of feedback to provide accurate and fast movements. The reason why the emergence of robots is relatively recent lies in the evolution of fast, cheap and reliable computers which form the heart of the robot and provide both control and reprogrammability.

The availability of cheap computing has also meant a parallel development of sensors and sensory processing. This, with the inherent flexibility provided by having the robot's motions commandable from a controlling computer, means that the robot's movements can be modified according to the sensory information. If, for example, the position of a certain component is known only inexactly, a camera can be positioned over the component and a more exact position extracted from the camera data by a sensory processor. The position is transmitted to the robot which then picks up the component correctly. This is shown schematically in figure 1.2.

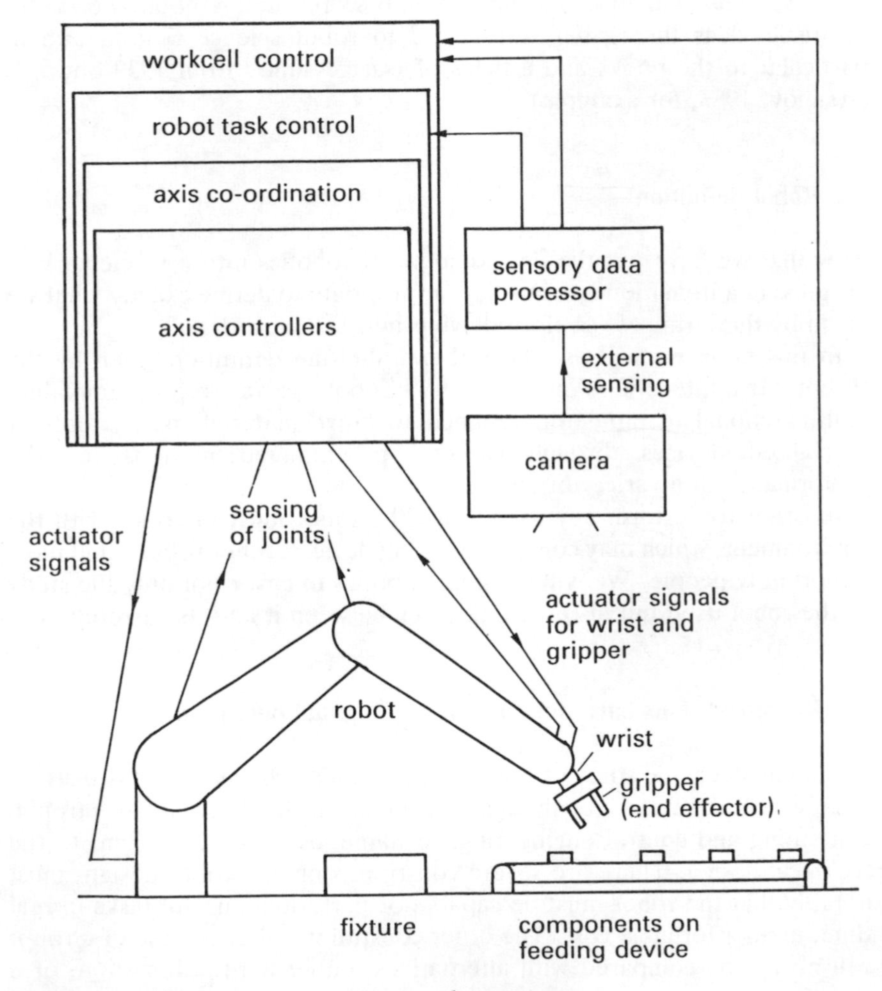


Figure 1.2 Schematic of a sensory robot system

If the camera is mounted on the end effector of the robot, it can continuously monitor the error between the component position and the end effector. This arrangement has several advantages over that shown in figure 1.2, as will be discussed in chapter 6.

However, although these machines are much more adaptable than either simple or reprogrammable sequence machines, their reactions to sensory feedback must be programmed explicitly. They will react to an error condition, in the above case a poorly positioned component, only if this has been foreseen by the programmer. Research workers are now studying the application of artificial intelligence techniques so that such explicit definition of error recovery actions is no longer necessary. The robot will have the intelligence to decide what it should do, or even what it should try to sense. This will form the limit of the scope of this book. For more futuristic ideas the reader is referred to robot science fiction, and in particular to the books and articles of Isaac Asimov from 1939 onwards (Asimov, 1983, for example).

## 1.2 Robot definition

Now that we have put the development of robotics into a framework of progress in automation systems, it is appropriate to define exactly what we mean by the term *robot* before delving into technical detail.

In this book we will use the rather utilitarian definition given by the Robot Institute of America: "A robot is a reprogrammable, multifunctional manipulator designed to move material, parts, tools or specialised devices through variable programmed motions for the performance of a variety of tasks."

In order to perform any useful task the robot must interface with the environment, which may comprise feeding devices, other robots, and most importantly people. We will consider robotics to cover not only the study of the robot itself but also the interfaces between it and its surroundings.

## 1.3 Robotics versus hard automation and manual operation

To an engineer, the study of robotics offers many challenges. It is truly an area which covers many disciplines: mechanical, electrical, electronic, computing and control engineering, to name just those important to the technical design. There are severe constraints on any robot design, most notably that the robot must be capable of performing useful tasks in real time, at an affordable cost. The other constraint is that the use of a robot will always be compared with alternatives, either hard automation, or a human.

Hard automation can be defined as the use of dedicated pieces of machinery, typically to produce the same item over long periods of time. This is expensive to build and inherently inflexible to product changes. However, its design can be optimised to produce the maximum amount of product at a minimum cost, so it is more attractive than the use of robots for the large-scale production of a few different items.

However, much of industry is concerned with batch production where perhaps one type of item is made during the morning and another during the afternoon. Human beings are very good in this environment. From a robotic point of view they are light, mobile structures with exceptionally good sensory perception and an intelligence far above that of any current robot. This gives them superb adaptability. However they tire, may become unreliable, unpredictable, and may well wish to be pursuing other activities which give greater scope for the use of their intelligence, or indeed just give greater pleasure.

A robot will neither be optimised for a particular application nor have the adaptability of the human. However, it can combine the reliability and predictability (at least until robots are made 'intelligent') of hard automation systems with a little of the adaptability of the human. Robots therefore have a place somewhere between these two extremes. For robots to play a positive part in supporting human activity, not only must they adequately perform a given task but the human aspects of any implementation must be thoroughly considered.

As stated by Rosenbrock (1979): "Any alternative technology should surely make better use of human abilities. It should not use men and women to perform meaningless, fragmented jobs which reduce them to automata. But this is not the same thing as suggesting a return to a more primitive craftsmanship, however satisfactory that may have been in its day. The problem is rather to use the best technology that we know, but to make it an aid to those who work with it, so that their work becomes an enrichment, not an impoverishment of their humanity, and so that the resource which their abilities represent is used to the highest degree."

Unfortunately there is insufficient space in this book to develop this and related economic factors.

In the justification of a robot installation, the easiest approach is to compare the local costs of the manual, robotic and hard-automation alternatives. But the values of any measures which are used will vary greatly according to the boundaries within which they are calculated. Let us assume that it is proposed that a particular subassembly operation is to be carried out by a robot instead of a person. If the boundary is drawn tightly round this particular operation, then the economic effects can be quantified in terms of production costs per unit, allowing for running costs, depreciation and overheads. If the complete product is brought inside the boundary, there may be increased cost benefits at later stages in the robotic assembly because our particular subassembly is better made. If the boundary is further widened to include field service, repairs and maintenance, more benefits may be seen to accrue. If the customer is considered, a better-made product may be safer, or give better customer satisfaction.

Let us now include the operator. The manual subassembly job may be of poor quality in terms of human job satisfaction and a better-quality occupation might be available elsewhere. The opposite, of course, may be true. Note that the term 'occupation' includes employment or leisure. If all of the factory employees are considered, then the measures might change markedly. The loss of a single job may preserve the survival of many others in an intensively competitive market. When the complete market, or the whole country, or the world, is included in the boundary then the measures change yet again.

In short, any economic measures of the viability of a robot installation depend on the standpoint and the field of view of the person making the judgements. It is important not to be too simplistic and narrow-visioned in one's approach to economic justification.

It is appropriate now to return to the more technical matters concerning robot installations — but the above factors should be borne in mind, particularly in the design of the workcell and the man–machine interface. First of all, we will take a quick look at the types of task which might be undertaken and then consider some possible performance measures which might be used to select or design a robot for each case. This will enable us to see in a broad sense the elements which will be required to make the overall system. These elements will then be studied in more detail in later chapters.

## 1.4 Application requirements

The key element about a robot which differentiates it from an automation system is its ability to be reprogrammed. This gives enormous potential flexibility so that the task being undertaken may be changed with perhaps only a small change in the robot's program. Most important, this flexibility can be used to cater for uncertainties in the environment, provided that appropriate sensory devices are connected and the information from them is used to modify the robot's behaviour accordingly.

From an applications point of view, it is convenient to split up tasks into three types for which different types of movement control are needed.

### 1.4.1 Types of robot tasks

A very common operation in industry is that of *pick and place*. This would be typified by a packing operation, such as that shown in figure 1.3, where items are taken from a conveyor and placed into a pallet.

The position of the item on the conveyor and the pallet position provide constraints on the endpoints of the trajectory of the item, but it does not matter from a functional point of view what path is taken by the robot. In practice, the path which gives the shortest time to perform the task will be the best. Note, however, that it will be desirable to constrain the robot to avoid collisions. Thus, once it has gripped the object it should move upwards from the conveyor, then move across to a position over the pallet, and then move downwards before releasing the object.

The requirements for this operation are therefore the ability to specify start and end positions and to have movement through intermediate points. The performance of the task will be measured by the accuracy of the placement of the object and the overall cycle time for the operation. Note that this scenario also covers simple assembly tasks.

A different type of control is needed in applications such as arc welding or paint spraying. Here the object is a welding torch or spray gun, and its

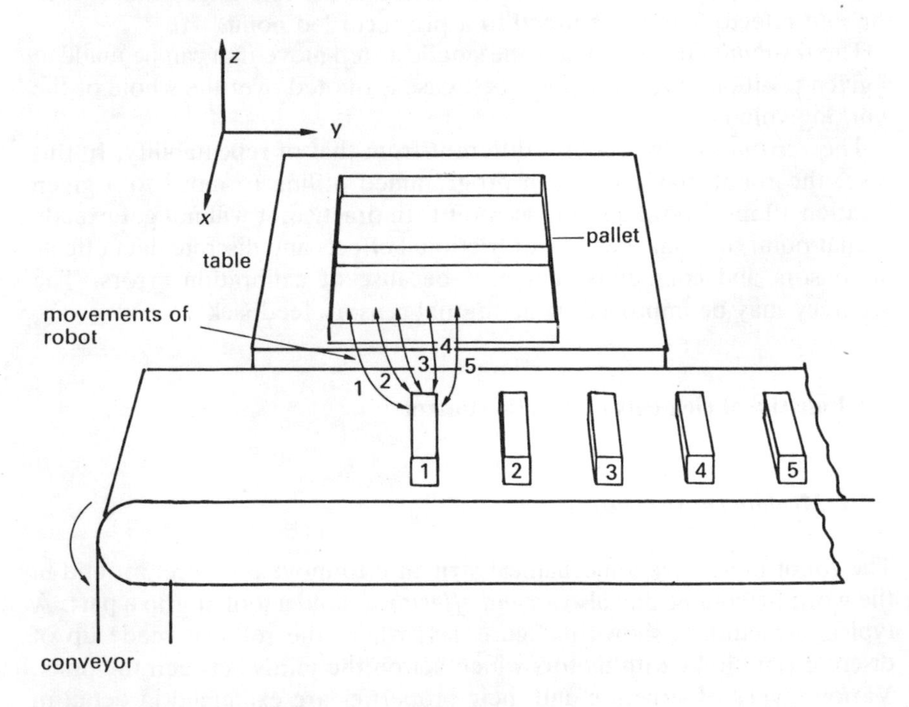


Figure 1.3 Pick-and-place operation

position and velocity need to be controlled at all points along its trajectory while it is in operation. This type of control is termed *continuous path control*.

A *sensory robot* may have the capabilities of either one of the above types, but will have the additional feature of being able to modify its target positions according to feedback from sensors. In the case of the pick-and-place robot, sensors could be used to detect the position of the object before pickup. In the continuous path case, they might be used to track the desired weld path ahead of the welding torch.

### *1.4.2 Measures of performance*

The *working volume* of the robot obviously needs to be sufficient so that all parts of the working area can be reached.

The *speed* and *acceleration* of the robot must be large enough so that the task can be accomplished within an acceptable time. Maximum speed is only a guideline, particularly when small movements are being made. In this case, the accelerations will dominate the time to accomplish the movement.

The *repeatability* of the robot is a measure of the tolerance within which the end effector can be returned to a pre-recorded point.

The *resolution* of the robot is the smallest step move that can be made at a given position. Typically, the worst case is quoted over the whole of the working volume.

The *accuracy* of the robot is different from that of repeatability. In this case, the robot may have been programmed offline to move to a given location 10 mm above a reference point. In practice, it will not get exactly to that point, not only because of frictional effects and discrete data effects in sensors and computers, but also because of calibration errors. The accuracy may be improved by additional sensory feedback.

## 1.5 Functional elements of robotic control

### *1.5.1 Mechanical structure*

The robot must have a mechanical structure to move an object around in the working volume and also an *end effector* to hold a tool or grip a part. A typical structure is shown in figure 1.4, where the robot is made up of discrete rigid links with motors which act on the joints between the links. Various types of structure and their properties are examined in detail in chapter 2.

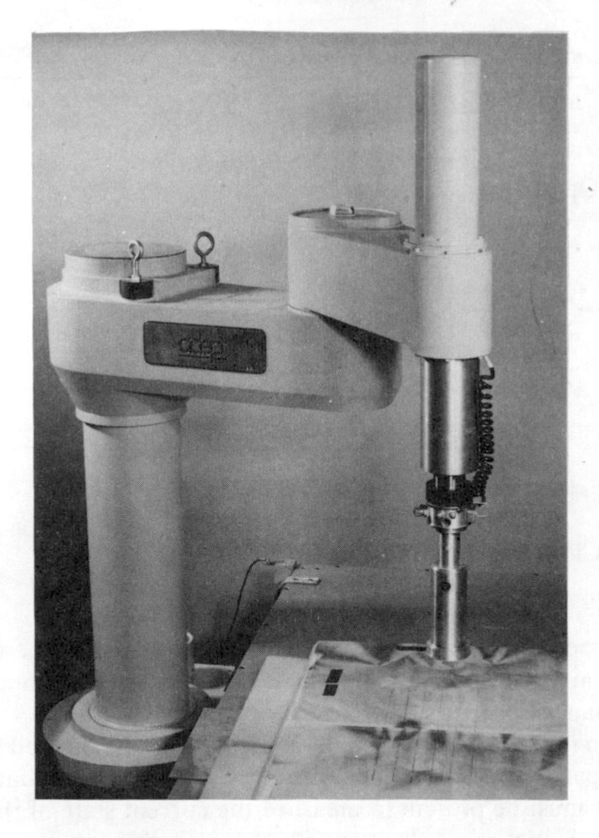

Figure 1.4 Adept I robot

### *1.5.2 Control of movements*

Given a particular structure, it will be necessary to determine the required movements of each part of the mechanism in order that the end effector can be moved to a required position and orientation in space.

Two types of control may be used in different parts of the robot, often depending on the type of actuators used. Open loop control is shown schematically in figure 1.5. No measurements are taken of the output position, so this control relies wholly on good calibration and the reliable behaviour of the actuator system.

In closed loop control, depicted in figure 1.6, the current output position is sensed and fed back to give an error signal which is used to drive the actuators.

The control requirements will essentially be the satisfaction of the performance requirements listed above. In addition, these performance measures must be achieved under different operating conditions, such as

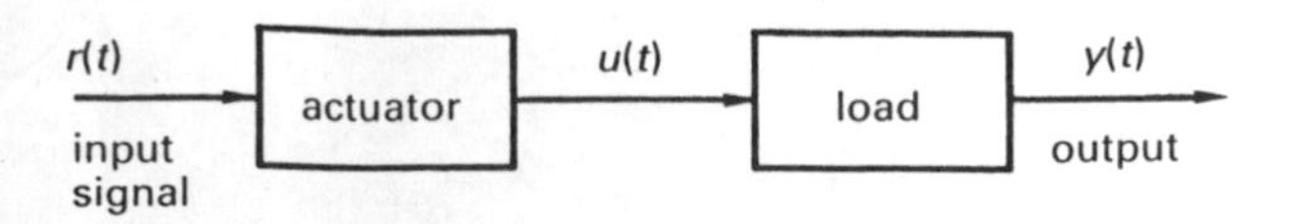


Figure 1.5 Open loop control

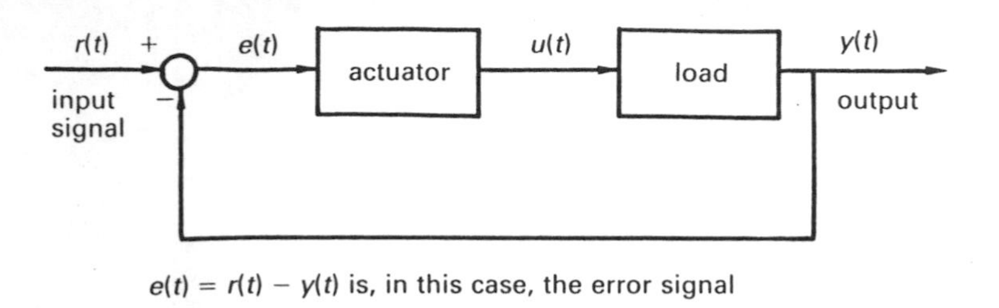


$e(t) = r(t) - y(t)$ is, in this case, the error signal

Figure 1.6 Closed loop control

change of load, variations in the mechanical behaviour of the robot, such as changes in friction, and variations in environmental conditions, for example, changes in temperature.

In order to realise the control system, actuators must be used to drive the elementary parts of the mechanical structure. If closed loop control is used, then sensors must be present to measure the current state of the positions of the joints. Sometimes velocity and acceleration signals are also directly measured. Details of commonly used actuators and sensors are given in chapter 3.

The requirements for high speeds and accelerations, high accuracies, and perhaps continuous path control, require that the dynamic behaviour of the robot be studied. It will be seen in chapter 4 how control techniques can be utilised to meet these needs.

### *1.5.3 Task control*

The task to be performed must be described in some manner and stored in the controller which is going to co-ordinate all the actuator movements. The positions may be taught 'online' or programmed 'offline'. Objects must be located and be manufactured to a precision which is sufficient to enable the task to be carried out reliably. Chamfering of holes or compliant gripping structures may be used to allow for some uncertainties.

The different methods of programming the robot together with the associated man–machine interfaces are described in chapter 5.

Chapter 6 discusses the implications of sensory feedback, with a description of some sensors and data processing techniques. Sensory feedback also affects the task control language required and the programming effort needed. Some tasks, such as the fabric-handling problem described in chapter 7, cannot be undertaken with the required accuracy and reliability without the use of sensory information. In such cases, sensory feedback is extending the range of practical applications of robotics.

### *1.5.4 Workcell control*

The robot, with or without its sensory systems, may be regarded as part of a workcell with associated devices and machines such as conveyors. An individual workcell may be just one part of a larger system. This may, for example, be a complete manufacturing plant. There will then be a need to co-ordinate the activities of the workcells, even to the extent of being able to reschedule tasks between them. The manufacturing plant may itself be part of a *Computer Integrated Manufacturing* (CIM) system and be linked through to designers via a *Computer Aided Design Computer Aided Manufacture* (CADCAM) system. The consequent integration problems are considered in chapter 7.

Because of the fascination of moving, human-like machines, it is tempting to regard the robot as being the central and most important part of a workcell. It can perhaps be more properly considered as a very flexible manipulative device which is an integral part of a larger robotic system. This should be borne in mind throughout the earlier chapters which concentrate on the robot itself. Later chapters consider the robotic system as a whole.

# 2 Kinematics and Dynamics

The essential purpose of a robot is to move objects or tools from one position in space to another. This chapter considers the types of mechanical structure which can be used, and the various forms of end effector which are suitable for holding the object or tool. The types of actuator which can be used to move each part of the structure are described in chapter 3.

In order to control the movements of different parts of the structure, it is necessary to study the kinematic equations. A commanded movement defined in one set of co-ordinates, such as 'move the object vertically upwards by 3 mm, keeping its orientation constant', will have to be transformed into a movement in joint co-ordinates for a specific robot structure, giving a command such as 'rotate joint 1 by 30 degrees clockwise, extend link 2 by 1 mm, and rotate joint 3 by 30 degrees anticlockwise'. It is also important to be able to make a transformation in the opposite direction – that is, to determine how the object moves in the world when the joints are moved by certain amounts. The dynamic behaviour of the robot must also be studied. Several performance measures are related to how the end effector moves as a function of time, so we must find how these are affected by the forces and torques applied by the actuators.

## 2.1 Degrees of freedom

Six parameters are necessary and sufficient to define the position and orientation of a general object in three-dimensional (3D) space. The position of the object can be defined in Cartesian co-ordinates ($x$, $y$, $z$) relative to some fixed reference point. Alternatives are to define the position in cylindrical or spherical co-ordinates. These schemes are shown in figure 2.1.

Using Cartesian co-ordinates the orientation can then be defined by a sequence of three rotations about the $x$, $y$ and $z$ axes. If the same terms are used as for a ship or aircraft, then the rotations corresponding to *roll*, *pitch* and *yaw* are shown in figure 2.2.

Orientation may also be defined by the *Euler angles*, depicted in figure 2.3. If a set of rectangular co-ordinate axes is fixed to an object, its orientation may be expressed as a succession of rotations about each axis.

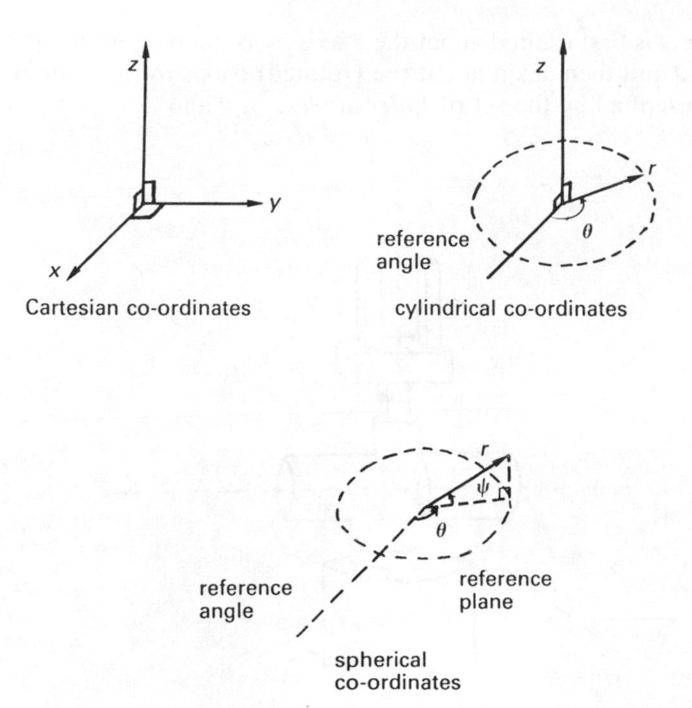


Figure 2.1 Co-ordinate frames

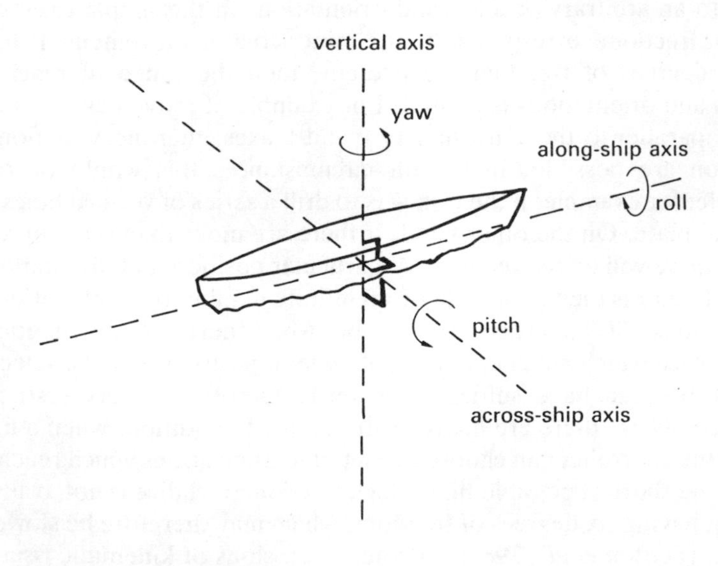


Figure 2.2 Roll, pitch and yaw

If the object is first rotated about the $z$-axis by $\phi$, then about the (rotated) $y$-axis by $\theta$ and then again about the (rotated) $z$-axis by $\psi$, its orientation may be described by the set of *Euler angles*, $\phi$, $\theta$ and $\psi$.

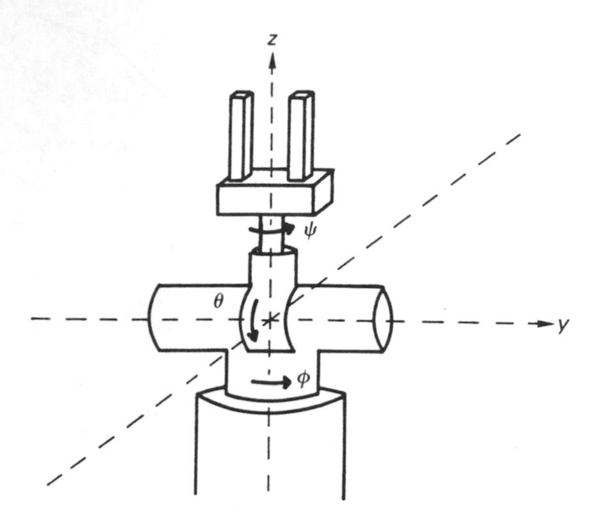


Figure 2.3 Euler angles $\phi$, $\theta$ and $\psi$

A robot therefore needs six degrees of freedom if it is to move the end effector to an arbitrary position and orientation. In the simple case, each degree of freedom corresponds to one joint actuator movement. If fewer than six degrees of freedom are present, then the range of reachable positions and orientations is limited. For example, if movements can only be made parallel to the Cartesian $x$, $y$ and $z$ axes, then no variations in orientation are possible. In certain circumstances this would be quite acceptable: for example, if the task was to drill a series of vertical holes in a horizontal plate. On the other hand, if there are more than six degrees of freedom there will be *redundancy*. A particular position and orientation of the end effector is then attainable by an infinite number of combinations of joint positions. This is especially valuable when there are obstructions in the workplace which mean that alternative joint positions may be selected. Such a robot can have sufficient dexterity to work in very restrictive workspaces where there are many obstructions. In addition, when a move is made, the controller can choose the set of joint positions which reach the target in the shortest possible time. Such freedom of choice is not available in a robot having six degrees of freedom, which may therefore be slower in operation (Fenton *et al.*, 1983). Further discussions of kinematic issues in redundant robots are given by Wampler (1989).

## 2.2 Mobility

Robots are usually constructed from a series of rigid links connected by joints. The particular type of joint defines how one link can move relative to the other. However, there are alternatives to the serial link–joint–link approach and these will be explored in section 2.3.

### *2.2.1 Joints*

Two types of joint, *prismatic* and *revolute*, are commonplace. a prismatic joint, also known as a *sliding joint*, enables one link to be 'slid' in a straight line over another. A revolute joint, if we consider the one degree of freedom case, takes the form of a hinge between one link and the next.

Two or more such joints can be closely combined, as shown in figure 2.4.

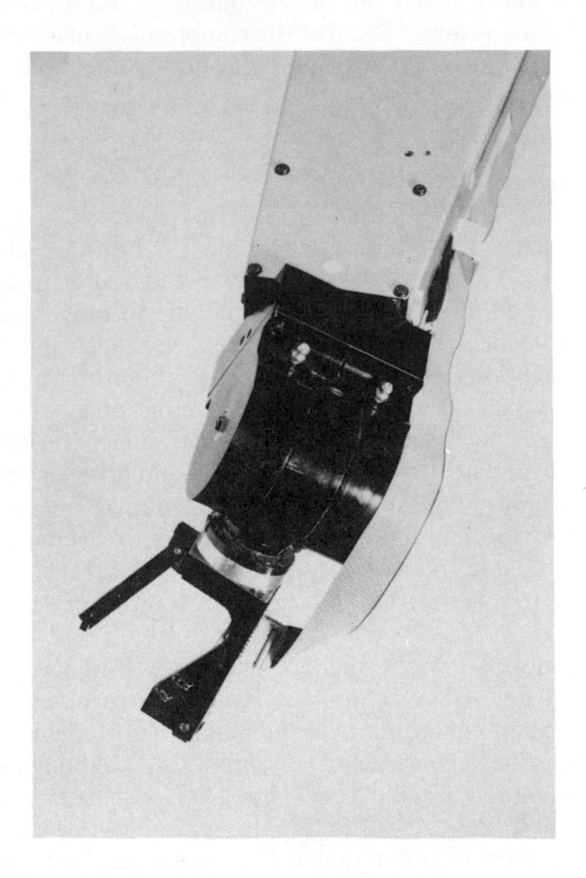

Figure 2.4 The Unimation PUMA robot's wrist

This illustrates the PUMA robot's wrist which has three degrees of freedom.

A ball and socket joint has the same effect as combining three simple revolute joints.

### 2.2.2 Links

In order to achieve the fastest possible response for a given movement and actuation system, the links forming the structure must be kept as light as possible. In order that deflections of the links under static and dynamic loads be kept to a minimum, the links must also be as rigid as possible. These two requirements conflict and a compromise must be reached.

In practice there are many more factors to consider such as cost, requirements for housing actuators, drive shafts and gearboxes, vibrational behaviour, non-elastic behaviour such as buckling, and the need to achieve a specified working volume. Some of the compromises made in practice are illustrated by the various photographs of robots in this book.

### 2.2.3 Mobile robots

Mobile robots have, by definition, some means of moving about, such as wheels or tracks. However, they will be considered here as fixed-base robots with the additional feature of being able to move the base. Such means include wheeled vehicles, tracked vehicles, and bipedal or multipedal locomotion. Mobility allows the robot to move from one workplace to another, or to move objects large distances without the need for a dedicated transportation system. In a more limited sense, conventional fixed-base robots have been mounted on overhead rails to give greater mobility, typically in a single direction.

Wheeled robots usually require a fairly flat floor over which to work. This is often available in factories and warehouses. Tracked vehicles can move about on rougher terrain, the famous Moon Buggy being one such example.

Legged locomotion devices, such as that shown in figure 2.5, are still in their infancy. They are very limited in performance, especially when compared with their counterparts in the animal world. Several multipedal mechanisms have been constructed (Fischetti, 1985) although some degree of human guidance is usually necessary. One particular problem being tackled (Choi and Song, 1988) is how to determine the sequence of leg movements required in order that a multipedal machine can overcome obstacles such as inclined planes, ditches and walls. At the far extreme of

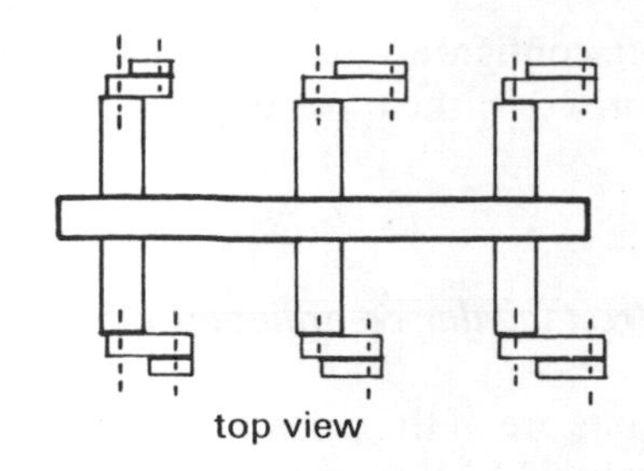


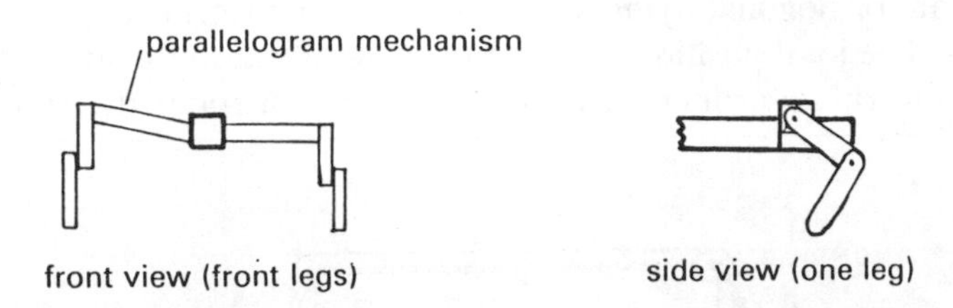


Figure 2.5 Multipedal mobile robot

legged locomotion is the one-legged hopping device successfully demonstrated by Raibert (1986).

## 2.3 Robot structures

Consider a robot having six degrees of freedom with five serial links connected through six joints. These joints could be either prismatic or revolute, giving $2^6$ possible combinations of prismatic and revolute joints. If we also take into account the possible orientations of the axes of each joint with respect to one of its attached links, then the number of possible joint–link configurations becomes infinite. As will be seen in section 2.3.5, parallel link configurations are also possible, as are the continuously flexible links discussed in section 2.3.6.

It is common to split the mechanical configuration of a serial robot into two. The first part comprises the links and joints giving the first three degrees of freedom starting out from the fixed base. These links provide the gross movement capability of the robot. The second part comprises the remaining degrees of freedom, as embodied by a 'wrist' such as the one shown in figure 2.4. The wrist is used to provide the means to alter the orientation of the end effector and usually consists of three revolute joints with intersecting axes of rotation. Note that for some applications, as noted above, a wrist is not necessary.

Some common configurations of serial links and joints will now be described, categorised by the types of joints giving the first three degrees of freedom.

### 2.3.1 Cartesian (rectangular co-ordinate) robots

The first three joints are of the prismatic type, giving a box-shaped working volume, as exemplified by the robot illustrated in figure 2.6. The first three joint axes are orthogonal, lying along the Cartesian world co-ordinate $x$, $y$ and $z$ axes. The joint positions are therefore identical to the end effector's position in world co-ordinates, making this type of robot the simplest to control.

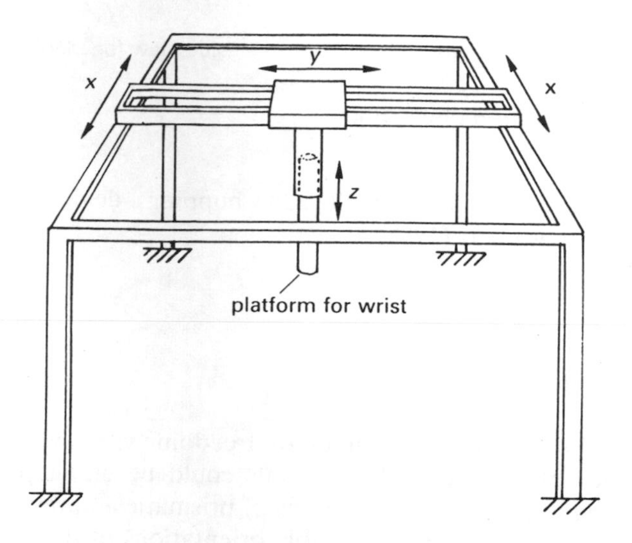


Figure 2.6 Cartesian co-ordinate robot

### 2.3.2 Revolute robots

All joints are of the revolute type. A typical example is shown in figure 2.7. This type of robot is often termed *anthropomorphic* because of the similarities between its structure and the human arm. The individual joints are named after their human-arm counterparts. These robots have a large working volume and are very popular. However, the control becomes far more complex than that of the Cartesian robot because the transformations required to go backwards and forwards between a vector of end effector positions in Cartesian co-ordinates and a vector of joint angles are now

Figure 2.7 A revolute robot (Unimation PUMA 560)

much more complicated, as is the dynamic analysis. This type of analysis will be described in sections 2.5 and 2.6.

### *2.3.3 Mixed revolute–prismatic robots*

The configuration shown in figure 2.8 is called a SCARA (Selectively Compliant Arm for Robot Assembly) type (Makino and Furuya, 1980). It can be made to have a high load-carrying capacity, and be rigid in the vertical plane and yet compliant in the horizontal plane. This can be advantageous for assembly tasks (see section 2.3.7). The key feature is the use of two or three revolute joints with vertical axes. Typically, a prismatic joint is also used, either just before or just after them. Control complexity lies between that for the Cartesian robot and the completely revolute robot.

### *2.3.4 Distributed robots*

Rather than using one fast robot having six degrees of freedom to perform a complex task, it may be more economic to split the task up into a series of operations carried out in parallel by simpler, cheaper robots. Indeed, some devices may have only one degree of freedom. Several manufacturers now offer robots in modular form. Individual axes of movement can be

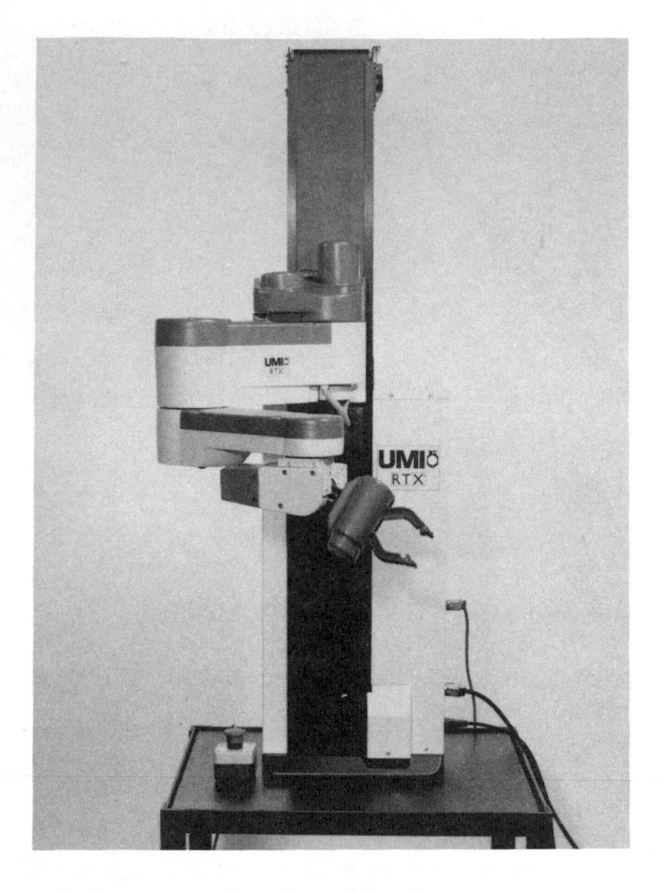


Figure 2.8 SCARA type robot (UMI-RTX)

purchased separately and combined to form any configuration desired by the user. For example, the simple pick-and-place task, shown in figure 1.3 using a single robot, could be performed by either a three-degree-of-freedom Cartesian robot, or by a two-degree-of-freedom Cartesian robot ($x$, $z$ axes) with the pallet sitting on a single-degree-of-freedom mechanism ($y$-axis).

### *2.3.5 Parallel robots*

Here the links are arranged in parallel rather than in series (Stewart, 1965). Figure 2.9 shows one parallel mechanism having six degrees of freedom. This has a small effective working volume for its size, but positioning errors are no longer cumulative as they are for a robot with serial links. The

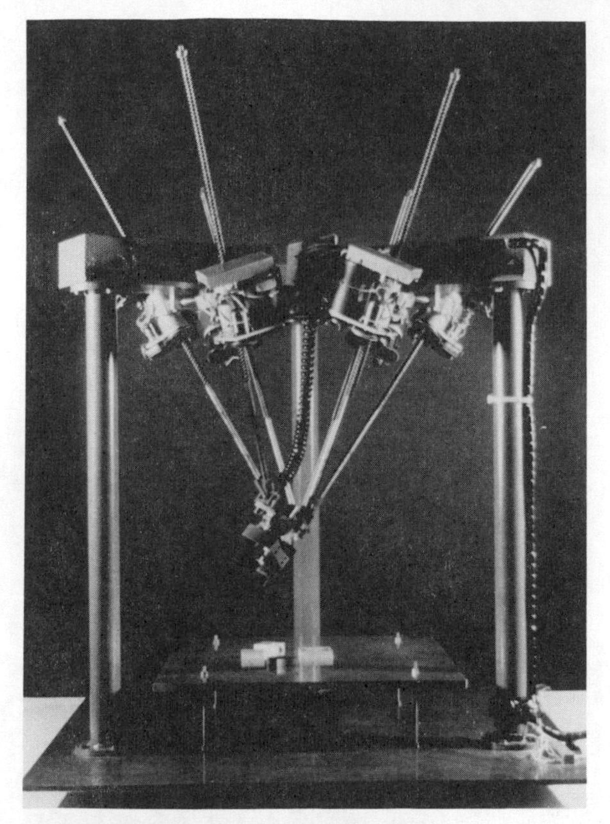

Figure 2.9 Parallel topology robot (GEC Gadfly) [courtesy GEC–Marconi Research Laboratories and Alvey 'Design to Product' project]

actuators are all mounted on the supporting frame so that they do not form part of the load as in many serially linked robots. The reduced load gives a potentially faster dynamic behaviour. An analysis of this structure is given by McCloy and Harris (1986).

The main disadvantage of purely parallel mechanisms is their restricted range of rotational movement. The GEC Tetrabot robot, illustrated in figure 2.10, overcomes this limitation by putting a wrist with three serial degrees of freedom on the end of a mechanism having three parallel links (Thornton, 1988).

### 2.3.6 Flexible link robots

All robot links will deflect to some extent under static and dynamic loads. Although it is usual to design the links to be as stiff as is needed, it is also

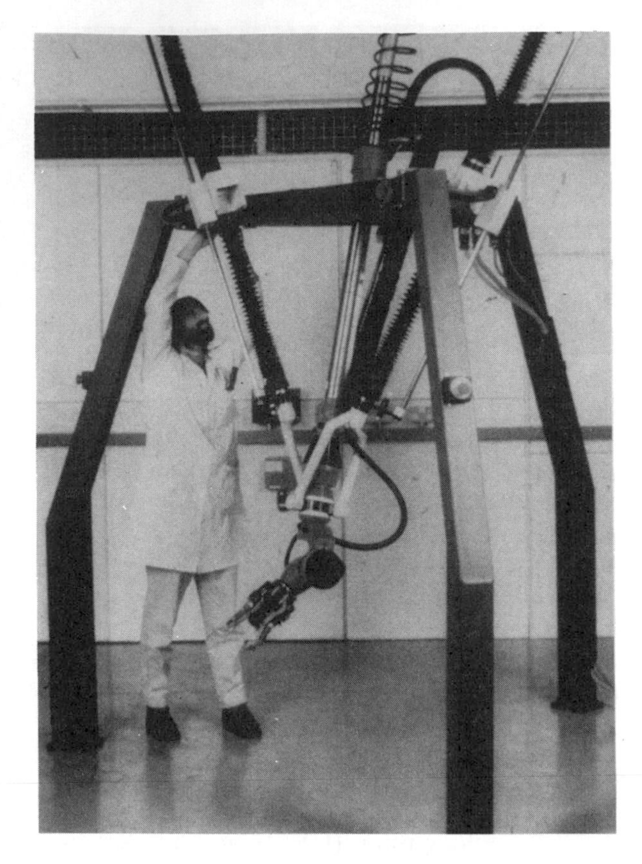

Figure 2.10 GEC Tetrabot robot [courtesy GEC–Marconi Research Laboratories and Alvey 'Design to Product' project]

possible to compensate for the flexibility by injecting extra signals into the actuators. The most common application for deliberately flexible links is in gripper finger design. The flexible fingers, shown in figure 2.13 with other gripping devices, will bend when a high air pressure is applied to them.

### *2.3.7 Compliance*

A non-sensory robot will move through approximately the same locations regardless of what is happening around it. Yet the objects on which the robot is working are not always of exactly the expected size and do not usually lie exactly in their expected positions. It is possible sometimes to cope with a degree of uncertainty without recourse to sensors. A particular technique, which has been used with notable success in automated assembly, is the use of compliance.

The basic design of the SCARA type robot, already shown in figure 2.8, gives a high stiffness in the $z$-axis. In the horizontal $x$–$y$ plane there may be significant springiness or *compliance*, particularly if there are long, elastic transmission systems between the actuators and the joints. The end effector may therefore be deflected elasticly in the $x$–$y$ plane by the load. Consider now the task, as depicted in figure 2.11, of inserting a shaft into a chamfered hole – a *peg-in-hole* task. If the shaft is slightly mislocated in the $x$–$y$ plane, then a robot without compliance will fail to insert the shaft. A robot having compliance allows the forces between the shaft and the chamfer to deflect the end effector in the horizontal plane.

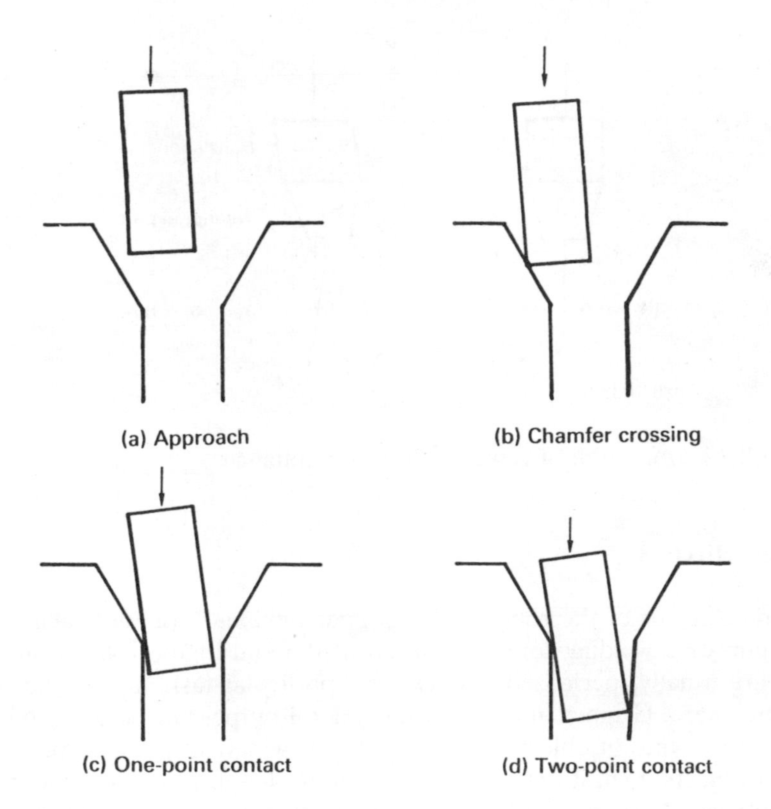


Figure 2.11 Peg-in-hole insertion sequence: (a) approach; (b) chamfer crossing; (c) one-point contact; (d) two-point contact

Let us now consider the case where the axes of shaft and hole are not parallel. In order to avoid wedging and jamming it is desirable that the shaft not only have lateral compliance (as described above) for the chamfer crossing and one-point-contact stages, but that it also has rotational compliance during the two point contact. The remote-centre compliance

(RCC) device (Whitney and Nevins, 1979) performs this function even if the robot is perfectly stiff. Figure 2.12 illustrates the principles of its operation.

The lateral compliance is provided by a sprung parallelogram mechanism. Rotational compliance is achieved using a second spring mechanism such that the centre of rotation is at the tip of the shaft. Both are combined into a unit which is attached to the wrist of the robot. A detailed mathematical analysis of the device and the forces during the insertion stages is given by Whitney (1986). Typical errors which the device can absorb are quoted as 1–2 mm and 1–2 degrees.

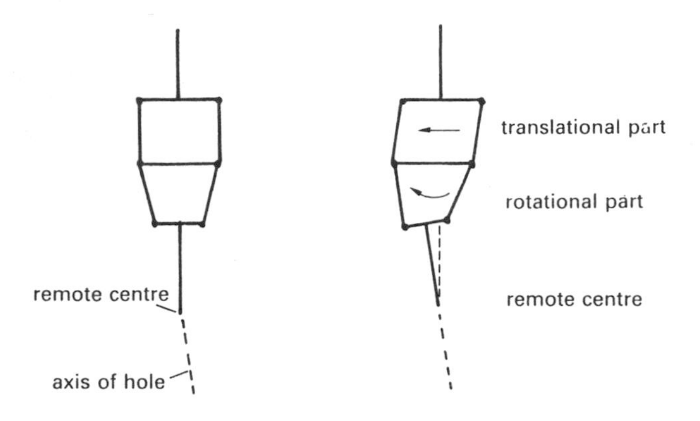


Figure 2.12 Operation of remote-centre compliance

## 2.4 End effectors

The end effector of the robot is the gripper or 'hand', or tool such as a spray gun or a welding torch, mounted at the end of the robot's 'arm'. Tools are usually specialised devices for a particular task and will not be described here. Grippers may be either general-purpose or be designed to hold certain types of object. If a task involves widely disparate types and sizes of objects, then it may not be possible to design a simple gripper to hold all parts. In such cases it can be beneficial to have a means of rapidly and reliably changing grippers. An alternative is to mount several grippers on a plate which can be rotated to offer the correct gripper up to the object.

A general-purpose gripper typically consists of two or more 'fingers' against which the object is held. A selection of mechanisms is shown in figure 2.13.

A survey of robot grippers and several gripper changing mechanisms is published by IFS Publications (Pham and Heginbotham, 1986).

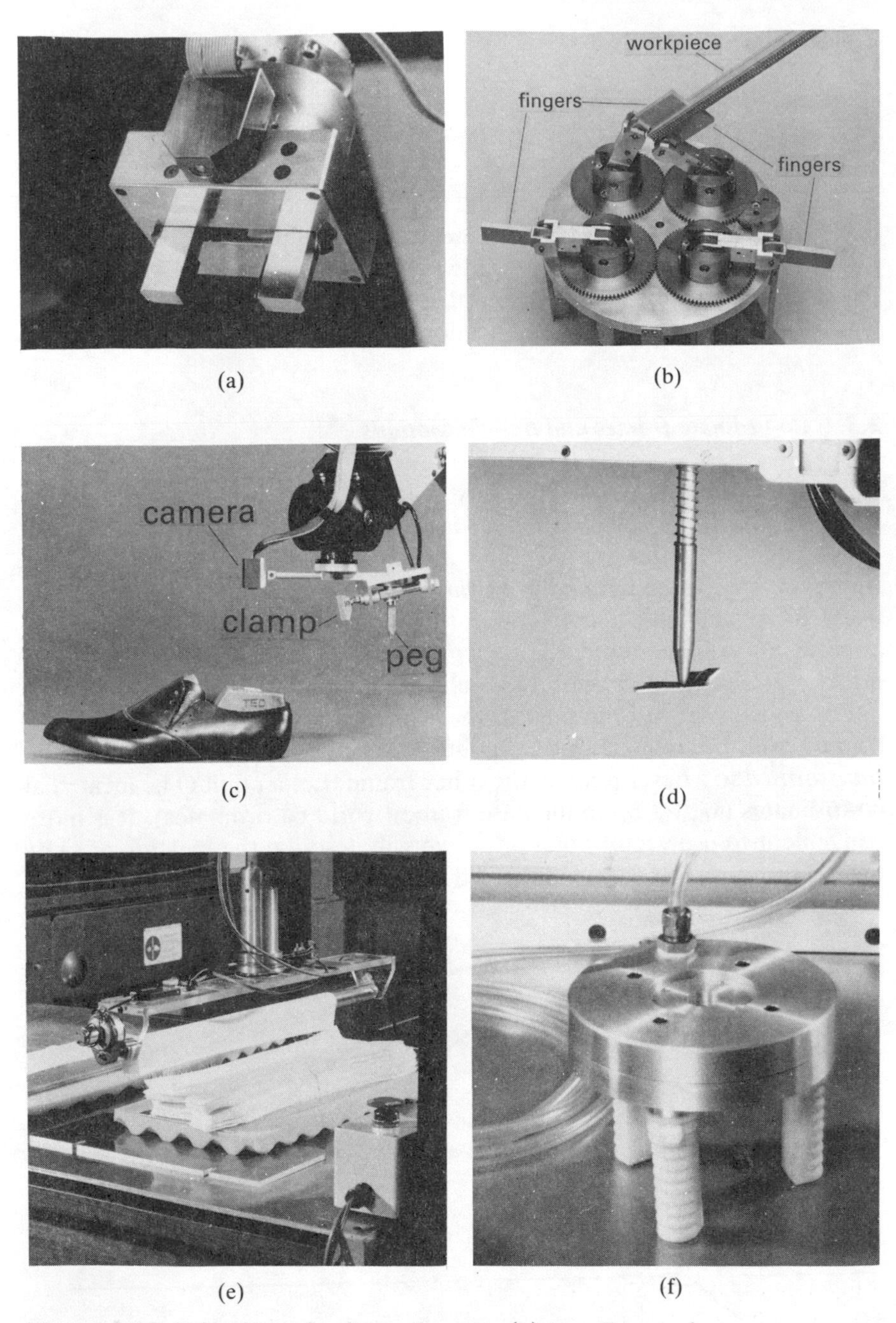


Figure 2.13 Selection of robot grippers: (a) two fingered; (b) multi-fingered; (c) peg-and-clamp; (d) vacuum; (e) electrostatic roller; (f) flexible fingers [part (b) courtesy of GEC–Marconi Research Laboratories and Alvey 'Design to Product' project]

## 2.5 Kinematic analysis

It is clearly necessary to know the location and orientation of the end effector given the states of all the joints. This is known as the *direct kinematics problem*. The *inverse kinematics problem* is to find the states of all the joints for a given location and orientation of the end effector. Note that 'state' in the above is taken to mean the angular position of a revolute joint or the displacement of a prismatic joint. As will be seen, we may want to use different co-ordinate systems to describe a movement or position. It will be assumed throughout this section that the links are perfectly rigid.

### *2.5.1 Co-ordinate frames and transformations*

We can describe the position of any point in space with respect to some arbitrary fixed co-ordinate system or *frame* in space. If the co-ordinate frame is fixed to the ground, then the co-ordinates of the point in this frame are said to be defined in *world co-ordinates*. For convenience, it is usual to fix this co-ordinate frame to the base of the robot. However, another co-ordinate frame, the *end effector frame*, which is fixed to the end effector as shown in figure 2.14, could be used equally well.

For a given set of joint angles, the world and end effector co-ordinate frames will be related and a point described in one frame can be *transformed* to a description in the other frame. Let a point Q be located at co-ordinates $(x_0, y_0, z_0)$ in the base frame (world co-ordinates). It is more convenient to use vector notation so we will describe the location of Q by the vector $\boldsymbol{q}_0 = [x_0, y_0, z_0]^T$ in world co-ordinates. Similarly, Q could be

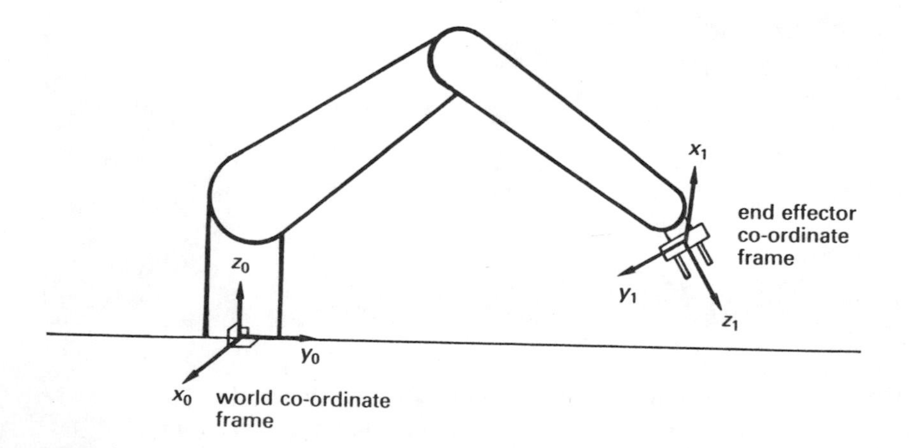


Figure 2.14 World and end effector co-ordinate frames

described by the vector $\boldsymbol{q}_1 = [x_1, y_1, z_1]^T$ in the end effector co-ordinate frame.

Appendix A gives the derivation of the elements of the transformation for two co-ordinate frames which have the same orientation but offset origins as shown in figure 2.15. This is followed by an analysis of the transformation between frames having coincident origins but disoriented axes, as seen in figure 2.16. Such orientation may be defined in terms of the Euler angles or the roll, pitch and yaw angles. It is also shown that the relationship between $\boldsymbol{q}_0$, a vector defining a point in co-ordinate frame 0, and $\boldsymbol{q}_1$, the description of the same point in co-ordinate frame 1, is given by

$$\boldsymbol{q}_0 = B\boldsymbol{q}_1 + \boldsymbol{p} \tag{2.1}$$

where $B$ is the appropriate orientation transformation as described in appendix A and $\boldsymbol{p}$ is the vector, expressed in co-ordinate frame 0, which goes from the origin of co-ordinate frame 0 to the origin of co-ordinate frame 1.

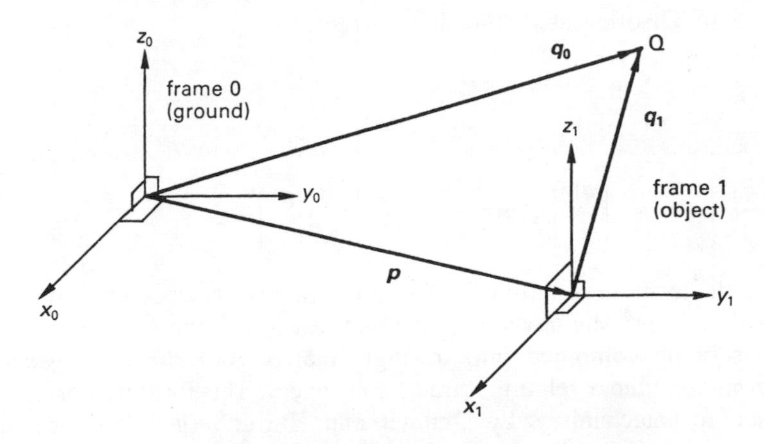


Figure 2.15 Co-ordinate frames with the same orientation

### 2.5.2 Homogeneous matrices

We have seen in the previous section how a vector in one rectangular co-ordinate frame may be transformed to a vector in another co-ordinate frame using equation (2.1). However, a much more convenient form for this equation, which will be used throughout the rest of the book, is obtained by augmenting the vectors $\boldsymbol{q}_0$ and $\boldsymbol{q}_1$ and the matrix $B$ to give

$$\boldsymbol{v}_0 = A_{01}\boldsymbol{v}_1 \tag{2.2}$$

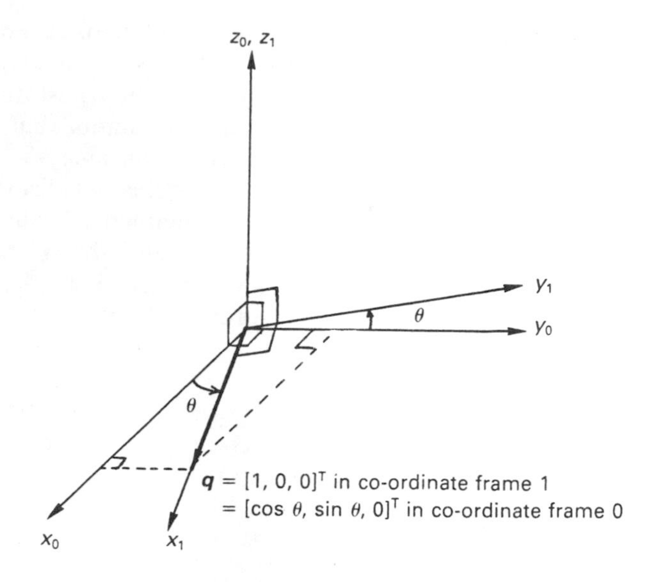


Figure 2.16 Disorientated co-ordinate frames

where

$$v_0 = \begin{bmatrix} q_0 \\ \hline 1 \end{bmatrix} \quad v_1 = \begin{bmatrix} q_1 \\ \hline 1 \end{bmatrix} \quad \text{and} \quad A_{01} = \left[\begin{array}{c|c} B & p \\ \hline 000 & 1 \end{array}\right]$$

When the terms in equation (2.2) are multiplied out, it becomes identical to equation (2.1). The advantage of this form lies in the translations and rotations being combined into a single matrix $A_{01}$, the *homogeneous transformation matrix* relating frame 1 to frame 0. This form was originally proposed for mechanisms by Denavit and Hartenberg (1955) but only became popular following the work of Paul (1981).

Note that, if the matrix $A_{01}$ is invertible, its inverse $A_{01}^{-1}$ can be used to pre-multiply both sides of equation (2.2) in order that $v_1$ may be found from a given $v_0$.

Let us now fix rectangular co-ordinate frame 0 to the base of the robot, frame 1 to the end of link 1, frame 2 to the end of link 2, etc. Since

$$v_0 = A_{01}v_1, \quad v_1 = A_{12}v_2, \quad v_2 = A_{23}v_3$$

$$v_3 = A_{34}v_4, \quad v_4 = A_{45}v_5, \quad v_5 = A_{56}v_6$$

then

$$v_0 = A_{01}A_{12}A_{23}A_{34}A_{45}A_{56}v_6 \tag{2.3}$$

Therefore by computing the $A$ matrices for each pair of frames attached to adjacent links, the *total transformation matrix* $T$ may be calculated as

$$T = A_{01}A_{12}A_{23}A_{34}A_{45}A_{56} \tag{2.4}$$

### 2.5.3 Joint transformations

The next step is to determine the $A$ matrix for a particular pair of links and a certain joint state (rotation or translation). An analysis is given in appendix B for revolute joints, but prismatic joints may be analysed in a similar manner (Paul, 1981). Referring to figure 2.17, the $A$ matrix for a revolute joint is defined by

$$A_{i-1,\,i} = \begin{bmatrix} \cos\theta_i & -\cos\alpha_i \sin\theta_i & \sin\alpha_i \sin\theta_i & a_i \cos\theta_i \\ \sin\theta_i & \cos\alpha_i \cos\theta_i & -\sin\alpha_i \cos\theta_i & a_i \sin\theta_i \\ 0 & \sin\alpha_i & \cos\alpha_i & d_i \\ 0 & 0 & 0 & 1 \end{bmatrix}$$

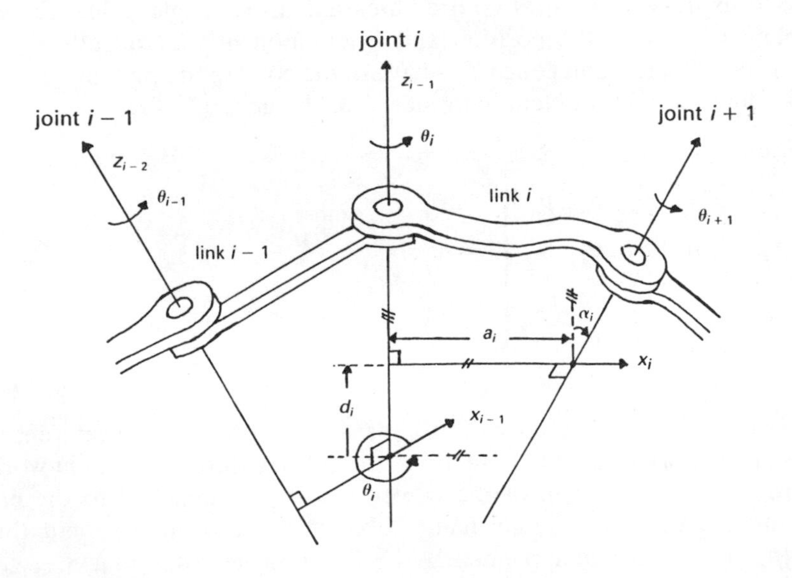


Figure 2.17 A general revolute joint–link structure

and for a prismatic joint the A matrix becomes

$$A_{i-1,\,i} = \begin{bmatrix} \cos\theta_i & -\cos\alpha_i \sin\theta_i & \sin\alpha_i \sin\theta_i & 0 \\ \sin\theta_i & \cos\alpha_i \cos\theta_i & -\sin\alpha_i \cos\theta_i & 0 \\ 0 & \sin\alpha_i & \cos\alpha_i & d_i \\ 0 & 0 & 0 & 1 \end{bmatrix}$$

A full kinematic analysis may now be carried out for a complete robot. The transformation matrices are derived for each joint and their product taken to produce the total transformation matrix as outlined in equation (2.4). An example is given in appendix B.

### 2.5.4 Solving kinematic equations

The previous sections have shown how the total transformation matrix $T$ may be obtained given a set of joint angles. Two further problems will be addressed in this section. The first is — given a set of joint angles and hence a derivation of $T$, what is the position and orientation of the end effector expressed in world co-ordinates and Euler angles? The second problem is — given the position and orientation of the end effector in world co-ordinates, and hence $T$, what are the corresponding joint angles?

Tackling the first problem, consider $T$ of the general form

$$T = \begin{bmatrix} n_x & o_x & a_x & p_x \\ n_y & o_y & a_y & p_y \\ n_z & o_z & a_z & p_z \\ 0 & 0 & 0 & 1 \end{bmatrix}$$

The symbols $n$, $o$, $a$, $p$ are used for consistency with the cited references.

From the work described in section 2.5.1, the position $(x, y, z)$ in world co-ordinates of the origin of the co-ordinate frame attached to the end effector is given by the right-hand column of the $T$ matrix, and thus $\boldsymbol{p} = [p_x, p_y, p_z]^{\mathrm{T}}$. From appendix A, the Euler angles result in a $B_{\text{Euler } \phi\theta\psi}$ matrix formed from a series of rotations about the axes. Denoting $T_{\text{Euler } \phi\theta\psi}$ to be the homogenous version of $B_{\text{Euler } \phi\theta\psi}$, and $\mathrm{ROT}(z,\phi)$, $\mathrm{ROT}(y,\theta)$,

ROT($z,\psi$) to be the homogenous matrix equivalents of $B_{z\phi}$, $B_{y\theta}$, $B_{z\psi}$ respectively as defined in appendix A, then

$$T_{\text{Euler } \phi\theta\psi} = \text{ROT}(z,\phi)\ \text{ROT}(y,\theta)\ \text{ROT}(z,\psi)$$

The appropriate ROT matrices may be formed, substituted in the above and solved (Paul, 1981, pp.65–70) to give the following Euler angles:

$$\phi = \text{atan2}(a_y,\ a_x) \quad \text{or} \quad \phi = 180° + \text{atan2}(a_y,\ a_x)$$

$$\theta = \text{atan2}(a_x \cos\phi + a_y \sin\phi,\ a_z)$$

$$\psi = \text{atan2}(-n_x \sin\phi + n_y \cos\phi,\ -o_x \sin\phi + o_y \cos\phi)$$

where the function atan2($a,b$) is as used in Fortran and other programming languages and gives an unambiguous result for the angle whose tangent is ($a/b$).

Let us suppose that a target position for the end effector is defined in world co-ordinates and its orientation in terms of roll, pitch and yaw. The values of these parameters define the total transformation matrix $T$. Joint angle demands must be sent to the axis controllers in order to move the robot, so these angles must be calculated from the desired values of the world co-ordinates and the roll, pitch and yaw parameters. This is the second problem. The required relationships for the elbow manipulator of figure 2.18 are given in appendix B, where the values of the joint angles are given in terms of the elements of the $T$ matrix.

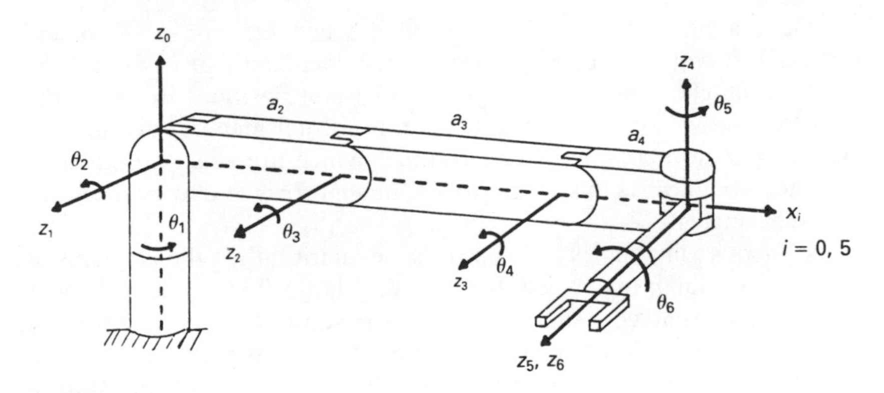


Figure 2.18 Elbow manipulator

### 2.5.5 The Jacobian

Let us suppose that the end effector of a revolute robot is moved a small distance ($\delta x$, $\delta y$, $\delta z$) in world co-ordinates. This will correspond to a set of changes ($\delta\theta_1$, $\delta\theta_2$, $\delta\theta_3$, $\delta\theta_4$, $\delta\theta_5$, $\delta\theta_6$) in the joint angles.

A *Jacobian* matrix $J$ may be formed, at this particular operating point in space, relating the changes in joint angles to changes in end effector position and orientation as given below:

$$\begin{bmatrix} \delta x \\ \delta y \\ \delta z \\ \delta\phi_x \\ \delta\phi_y \\ \delta\phi_z \end{bmatrix} = J \begin{bmatrix} \delta\theta_1 \\ \delta\theta_2 \\ \delta\theta_3 \\ \delta\theta_4 \\ \delta\theta_5 \\ \delta\theta_6 \end{bmatrix} \tag{2.5}$$

where $\delta\phi_i$ is a small rotation about the $i$th axis ($i = x, y$ *or* $z$).

Dividing both sides of the above equation by $\delta t$, the small time in which this action takes place, it can be seen that the Jacobian gives the relationship between the instantaneous velocities in world co-ordinates and in joint angles. A velocity demand described in world co-ordinates can thereby be transformed into a velocity demand in joint co-ordinates by pre-multiplying the world co-ordinate velocity demand vector by $J^{-1}$.

### 2.5.6 Singularities

To achieve a particular small change in the end effector position and orientation, it will be necessary to invert the Jacobian matrix to find the corresponding changes in joint angles. At certain positions in space, the matrix may prove to be non-invertible in the usual manner. A Jacobian matrix with a zero column or two columns which are exact multiples of each other would cause this to happen. A *singularity* is said to exist for this set of joint angles.

If the robot's joint angles are close to a singularity, then the elements of the inverse Jacobian matrix will become very large. Large angle changes are required to achieve a small change in the state of the end effector in world co-ordinates. Similarly, a requirement for a constant small speed in world co-ordinates will transform, via the inverse Jacobian, into a requirement for high joint speeds which may be physically unattainable. In addition, the elements of the inverse Jacobian matrix will vary quickly as the singularity is approached. The required joint speeds will therefore change rapidly for the movement defined above. Such rapid changes in joint speeds correspond to joint acceleration demands which may be unattainable by the actuation system.

The robot is likely to perform badly when close to singularities, and so these locations should be known and avoided during the task design.

## 2.6 Dynamics

The previous sections have dealt solely with the relationships between positions and velocities described in two alternative reference frames. Movements from one position to another will be effected by various actuators and it is necessary to study the movements as a function of time in order that design requirements, such as high speeds and no overshoot, may be achieved. Some method is therefore required to obtain the dynamic equations of the robot. When solved, these will show how the joint positions vary with time when the actuators are energised. In this work Lagrangian mechanics will be used, as the method is simple and can be used systematically for very complex structures.

The Lagrangian function $L$ may be defined as the difference between the kinetic energy $K$ and the potential energy $P$ of the system.

$$L = K - P$$

As an example, figure 2.19 shows a schematic of the two horizontal links of a SCARA type robot (see figures 1.4 and 2.8 for actual examples). If we assume that the masses of the links are negligible in comparison with the masses of the actuators located at the joints, and that the movements in all other joints have no dynamic effect on the two horizontal links, then the dynamic equations can be determined quite easily. From appendix C

$$F_1 = (m_1 + m_2)a_1^2\ddot{q}_1 + m_2a_1a_2(\ddot{q}_2 \cos q_2 - \dot{q}_2^2 \sin q_2) \\ + m_2a_2(a_2\ddot{q}_1 + za_1\ddot{q}_1\cos q_2 - za_1\dot{q}_1\dot{q}_2 \sin q_2 + a_2\ddot{q}_2)$$

and

$$F_2 = m_2a_2\{a_2(\ddot{q}_1 + \ddot{q}_2) + a_1\ddot{q}_1\cos q_2 + a_1^2\dot{q}_1^2 \sin q_2\}$$

The terms are defined in figure 2.19.

These non-linear ordinary second-order differential equations must be solved in order to obtain the equations of motion.

Note that the effective inertia of joint 2 as seen by its actuator is $F_2/\ddot{q}_2$. This changes with $q_1$ and $q_2$ and their derivatives. A simple linear control system would be designed to give optimum performance for a particular effective inertia. Changes in inertia, such as those seen in the simple case above, will cause the control to become non-optimal and so more sophisticated controllers are often used, as will be seen in chapter 4.

It can be shown (Paul, 1981) that if the energy dissipative elements, such as frictional effects, are neglected then the dynamic equations for an $n$-degree-of-freedom manipulator are given by

$$F_i = \sum_{j=1,n} D_{ij}\ddot{q}_j + \sum_{j=1,n}\sum_{k=1,n} D_{ijk}\dot{q}_j\dot{q}_k + D_i$$

where $F_i$ represents the force or torque applied to joint $i$

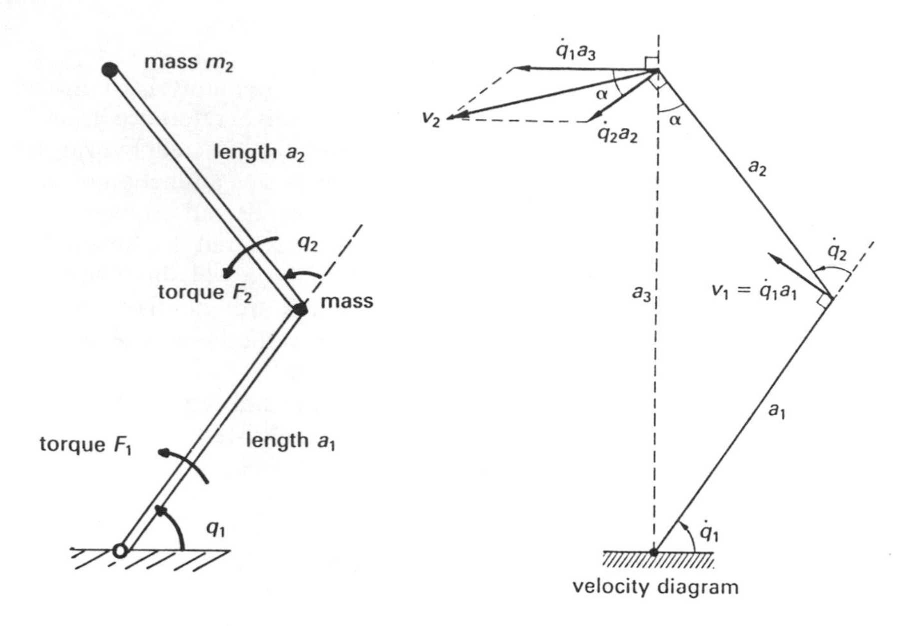


Figure 2.19 Dynamic variables for a two-link manipulator

$D_{ii}$ represents the effective inertia at joint $i$, including the actuator inertia
$D_{ij}$ represents the coupling inertia between joint $i$ and joint $j$
$D_{ijj}$ represents centripetal forces at joint $i$ due to a velocity at joint $j$
$D_{ijk}$ represents Coriolis forces at joint $i$ due to velocities at joints $j$ and $k$
$\mathrm{D}_i$ represents the gravity loading at joint $i$.

Note that if $D_{ij} = 0$ for all $j \neq i$, and $D_{ijk} = 0$, then the above equation reduces to a set of $n$ equations:

$$F_i = D_{ii} + D_i$$

and each axis may be controlled without reference to the dynamics of the other axes. Unfortunately, these terms cannot be neglected, particularly when high speeds are required. This is one of the reasons why the control of a robot can be such a challenge.

The above equations will be seen again in chapter 4, during the design of a robot control system. A robot will be controlled by sending signals to actuators and reading signals from sensors. These actuators and sensors have their own special properties which will affect the controller design. The next chapter is therefore devoted to them, with the kinematics and dynamics of the robot structure re-emerging in the control aspects studied in chaper 4.

# 3 Sensors and Actuators

This chapter describes the basic principles of pneumatic, hydraulic and electrical actuators, with special attention to those aspects which can limit the performance of a robot. A section on drive systems is included since the means of transmitting mechanical power from actuator to joint may also give rise to significant performance characteristics. The final sections describe methods of sensing the positions, velocities and accelerations of links. Such internal sensors are required to perform the dynamic control. External sensors, such as cameras, which are needed for task control, are described in chapter 6.

## 3.1 Hydraulic actuators

Hydraulic actuation systems comprise the actuator itself, which may give linear or rotary motion, the valves which control the flow of hydraulic fluid to it, and the supply which provides a source of high-pressure fluid.

### *3.1.1 The actuators*

Both single and double acting types, as shown in figure 3.1, are in common use. Rotary forms are also available. In the single-acting type there must always be a positive external force $F$ applied to return the piston after extension. The fluid in the cylinder will be (almost) incompressible and will be applied at pressures of up to 200 bar ($200 \times 10^5$ N/m$^2$, 2901 lb/in$^2$). A flowrate of 0.25 l/s ($0.25 \times 10^{-3}$ m$^3$/s, 15.26 in$^3$/s) at 200 bar gives a power output of 5 kW ($5 \times 10^3$ N m/s, 6.71 hp). In the double-acting device, high-pressure fluid may be applied to either side of the piston via the action of control valves.

Cleanliness is essential. The fluid, usually oil, must be as free as possible from air. Otherwise it becomes significantly compressible, and the performance of the actuator degrades. Anyone who has experienced the effects of air in the hydraulic fluid of a car braking system will surely remember this. Another critical factor is the state of the metallic seals between the moving piston rod and the stationary cylinder end. If these seals become worn, then fluid will leak into the atmosphere. A mist of fluid

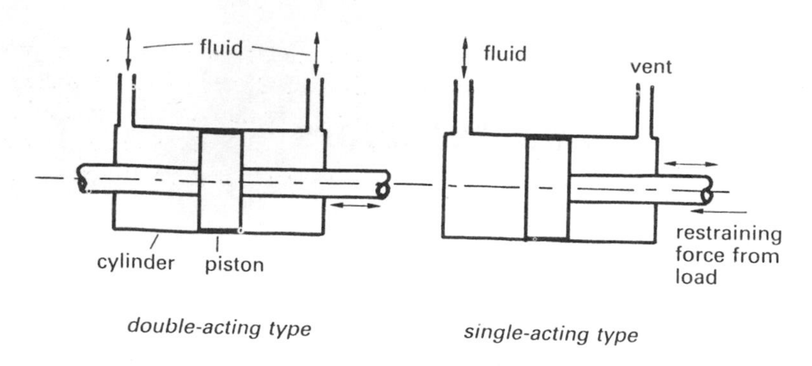


Figure 3.1 Hydraulic actuators

is produced or, in cases of more drastic failure, a jet of high-pressure fluid emerges. Regular maintenance should prevent the occurrence of the above problems.

### 3.1.2 Control valves

The spool valve, shown in figures 3.2, enables a small-movement, low-force input to cause a large pressure differential to be applied to the

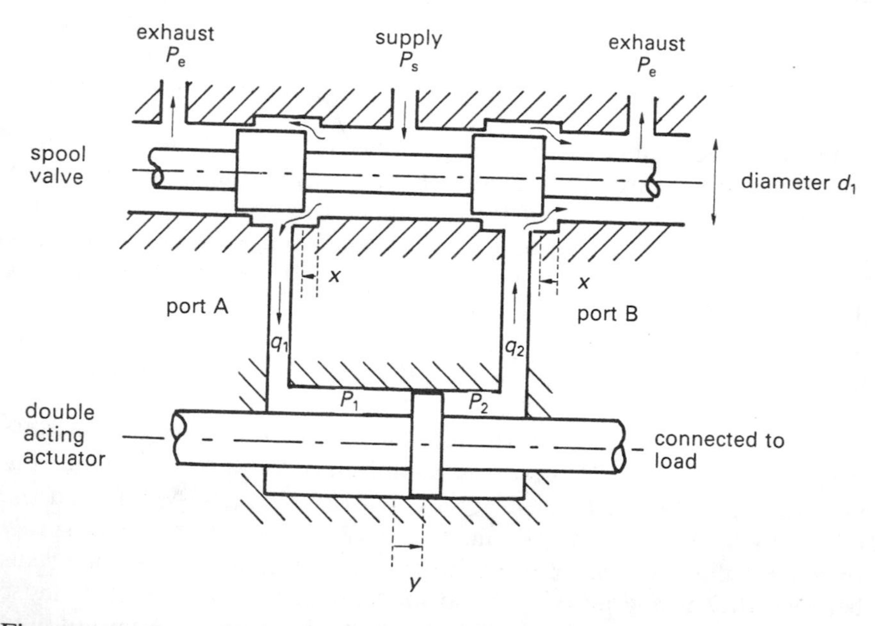


Figure 3.2 Hydraulic spool valve and actuator — after Stringer (1976)

hydraulic actuator, and hence a large force to be applied to the load. With the spool valve in the position shown in figure 3.2, port A is connected to the high-pressure side of the supply, and port B to the low-pressure side. A right-acting force will be applied to the piston. If the spool valve is moved $2x$ to the right, then the connections are reversed and a left-acting force results.

### 3.1.3 Hydraulic supplies

The power source is normally a constant-speed electric motor driving a hydraulic pump. The high-pressure oil is stored in a reservoir, also known as an accumulator, and then piped to the control valves. Problems are most likely to arise when all joints are suddenly, and simultaneously, commanded to run at full speed. If the accumulator is too small, or the electric motor has an inadequate power rating, the supply cannot supply energy at the rate demanded by the controller. The oil pressure drops, giving sluggish response.

### 3.1.4 Mathematical analysis

Before considering the full analysis of the four-way valve connected to the double-acting cylinder, as shown in figure 3.2, it is necessary to consider the flow of fluid through the annular orifices at the edges of the spool lands.

The volumetric flowrate $q_1$ through a sharp-edged orifice of diameter $d_1$ and width $x$ is given by the 'square root law':

$$q_1 = C_d \pi d_1 x \, (P_s - P_1)^{1/2} \left(\frac{2}{\rho}\right)^{1/2}$$

where the discharge coefficient $C_d \approx 5/8$ and the density of oil, $\rho \approx 870$ kg/m$^3$. Similarly, for the second orifice:

$$q_2 = C_d \pi d_1 x \, (P_2 - P_e)^{1/2} \left(\frac{2}{\rho}\right)^{1/2}$$

Assuming, for the moment, that $q_1 = q_2$ (and therefore $P_s - P_1 = P_2 - P_e$), $P_s$ is constant, $P_e$ is negligible, and setting $P_m = P_1 - P_2$, the equations can be simplified to

$$q = q_1 = q_2 = C_d \pi d_1 x \, (P_s - P_m)^{1/2} \left(\frac{1}{\rho}\right)^{1/2}$$

and then $q \approx 6.7 \pi d_1 x (P_s - P_m)^{1/2}$

Further simplifications, and linearisation of the equation gives (Stringer, 1976)

$$q = K_q x - K_c P_m \tag{3.1}$$

where the valve flow coefficient

$$K_q \approx 6.7\pi d_1 (P_s)^{1/2}$$

and the valve pressure coefficient

$$K_c \approx \frac{6.7\pi d_1 x (P_s)^{1/2}}{2P_s}$$

Note that these equations apply only when the pressures are measured in bars.

If we now consider the full valve–cylinder system, the volumetric flow of fluid into and out of the cylinder causes motion of the piston and compression of the fluid. Thus, if the piston is located centrally within the cylinder:

$$q = A\frac{dy}{dt} + \frac{V_t/2}{\beta}\frac{dP_1}{dt} = A\frac{dy}{dt} - \frac{V_t/2}{\beta}\frac{dP_2}{dt}$$

where $V_t$ is the total cylinder volume, $A$ is its cross-sectional area, and $\beta$ is the bulk modulus of the oil. Assuming, as before, that $q_1 = q_2$ and hence $dP_1/dt = -dP_2/dt$, then

$$q = A\frac{dy}{dt} + \frac{V_t}{4\beta}\frac{dP_m}{dt} \tag{3.2}$$

If the load has a mass $M$, with no frictional effects present or other external forces applied, then

$$AP_m = M\frac{d^2y}{dt^2} \tag{3.3}$$

Combining the above equations gives

$$\frac{K_q}{A}x = \frac{V_t}{4\beta}\frac{M}{A^2}\frac{d^3y}{dt^3} + K_c\frac{M}{A^2}\frac{d^2y}{dt^2} + \frac{dy}{dt}$$

Taking Laplace transforms gives the transfer function:

$$\frac{Y(s)}{X(s)} = \frac{K_q/A}{s\left\{\left(\frac{1}{\omega_n^2}\right)s^2 + \left(\frac{2\zeta}{\omega_n}\right)s + 1\right\}} \tag{3.4}$$

where $\omega_n^2 = \dfrac{4\beta}{V_t}\dfrac{A^2}{M}$

and $\dfrac{2\zeta}{\omega_n} = \dfrac{K_c M}{A^2}$

giving the damping ratio $\zeta$ as

$$\zeta = \frac{K_c}{A}\left(\frac{\beta M}{V_t}\right)^{1/2}$$

Figure 3.3 illustrates how the effective bulk modulus of the oil is reduced when air becomes trapped in it. From equation (3.4) it can be seen that this has the effect of reducing the natural frequency and hence slowing the response.

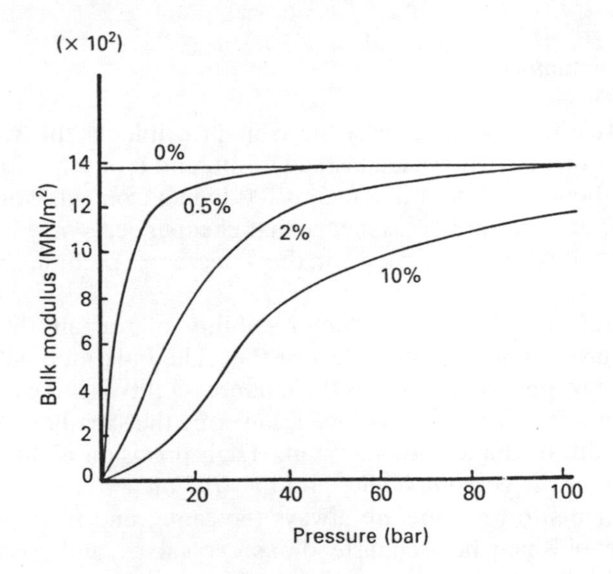


Figure 3.3 The effect of entrapped air on the bulk modulus of oil — after McCloy and Martin (1980)

If the fluid is assumed incompressible ($\beta = \infty$), then

$$\frac{Y(s)}{X(s)} = \frac{K_q/A}{s\left(\dfrac{K_c M s}{A^2} + 1\right)} \tag{3.5}$$

It has been shown (Williams, 1985) that the dynamic behaviour of hydraulic actuators can be critical to the dynamic performance of a hydraulic robot. It is therefore necessary to use such transfer functions of the actuators in the full model of the robot.

Note that the above analysis is for an ideal '*zero-lap*' or '*critical-centre*' spool valve. In this type, the spool lands *just* cover the orifices when in the central position. If there is overlap, there will be a *deadzone*; there will be no fluid flow for small movements of the spool from the central position. If the valve is of the *underlapped* type, there will be a leakage of fluid from the supply to the exhaust when the valve is centrally located. The deadzone is avoided but there is loss of energy and hence the dynamic performance is degraded. For details of these types of valve, and more details about hydraulic systems in general, the reader is referred to texts such as that by Stringer (1976) and McCloy and Martin (1980).

## 3.2 Pneumatic actuators

This type of actuator works on exactly the same principles as the hydraulic actuators of section 3.1. Air is used instead of oil, and typically this would be supplied at about 6 to 7 bar (about 85 to 100 lb/in$^2$) from a factory-wide supply. The use of lower pressures means that cheaper seals may be used, but smaller forces will be generated for the same cross-sectional area of cylinder.

The main drawback lies in the compressibility of the air. Study of equation (3.4) gives an idea of the effect of this. The bulk modulus is the reciprocal of the compressibility so, as the compressibility is increased, the undamped natural frequency $\omega_n$ is reduced, lowering the speed of response and the bandwidth of the actuation system. High precision is difficult to achieve because of the compressibility and the friction.

If the movements to be made are always the same, and from point to point, then solenoids may be used instead of servo valves, and precision is obtained by driving the piston against end stops. In this situation, pneumatic actuators are very common as they provide the cheapest effective solution. They are also frequently seen in end effectors. The gripping force may be easily adjusted by varying the air supply pressure.

## 3.3 Direct Current (DC) motors

Direct Current motors, of which there are many variants, are extensively used as actuators in robots. The simplest type is the permanent magnet motor which will be used to demonstrate the principles of operation and analysis. A major advance in recent years has been the emergence of solid state switching devices as part of the control circuitry.

### *3.3.1 Permanent magnet DC motors*

If an electric charge $q$ moves with vector velocity $\boldsymbol{v}$ through a magnetic field of intensity $\boldsymbol{B}$, it will experience a force $\boldsymbol{F}$ given by the vector cross product $\boldsymbol{F} = q\boldsymbol{v} \times \boldsymbol{B}$. Thus, if a current of $i$ amperes is passed through a conductor of length $l$ metres at right angles to the magnetic field of strength $B$ teslas, then a force is exerted on the conductor. This force will be at right angles to the plane containing $i$ and $B$, and will have magnitude $F = Bli$ newtons. This is known as the *motor law*.

The second important phenomenon is stated in the *generator law*. If a conductor of length $l$ metres is moved at a velocity of $v$ metres per second through a constant field of $B$ teslas, the directions of motion and field being at right angles to each other, then a voltage $e$ is generated across the conductor. The magnitude of $e$ is given by $e = Blv$ volts.

The basic construction of a typical permanent magnet DC motor is shown in figure 3.4. The field $B$ is stationary and is established by permanent magnets which form the *stator*. The turning part, the *rotor* or *armature*, comprises a set of coils of wire through which current is passed via brushes and a commutator. As the armature rotates, the commutator causes the brushes to contact different coils in turn. The coils and commutator are arranged such that the maximum torque is achieved as the current $i$ is passed through them. Both the motor and generator laws apply. The force on the conductors gives rise to a torque $T$ given by

$$T = K_{\mathrm{m}}i \qquad (3.6)$$

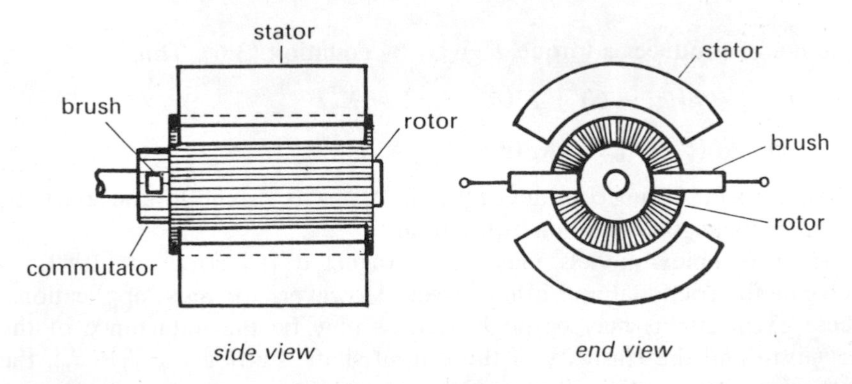


Figure 3.4 Permanent magnet DC motor

and the voltage $e$, generated as a result of the rotation, is

$$e = K_g\dot{\theta}_m \tag{3.7}$$

where $K_m$ is the *motor constant*, $K_g$ is the *generator constant* (note that $K_m = K_g$ if consistent units are used), and $\theta_m$ is the angular velocity of the rotor. The voltage $e$ opposes the current $i$. If an external voltage source of $v$ volts is applied across the armature which has windings of resistance $R$ and inductance $L$, then the electrically equivalent model of figure 3.5 results. Analysis of this circuit gives

$$L\frac{di}{dt} + Ri = v - K_g\dot{\theta}_m$$

If the mechanical side is now considered, the motor has a rotor inertia $J_m$ and is connected by a shaft, which may be elastic, to a load of inertia $J_l$.

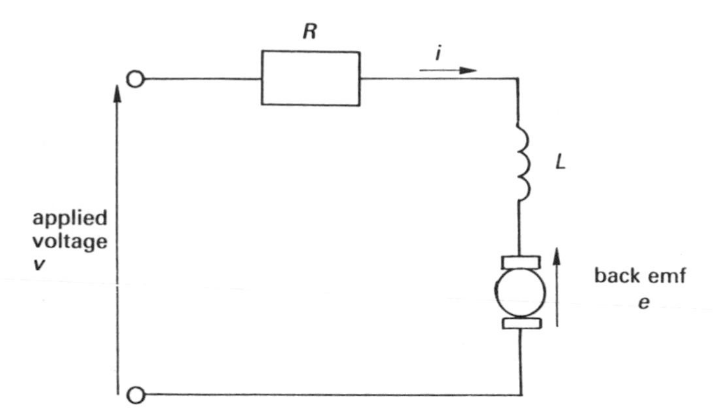


Figure 3.5 Equivalent circuit for a permanent magnet DC motor

The motor produces a torque $T$ given by equation (3.6). Thus

$$J_m\ddot{\theta}_m + b(\dot{\theta}_m - \dot{\theta}_l) + K_s(\theta_m - \theta_l) = K_m i$$

$$J_l\ddot{\theta}_l + b(\dot{\theta}_l - \dot{\theta}_m) + K_s(\theta_l - \theta_m) = 0$$

where $K_s$ is the spring constant of the shaft, $b$ is its viscous torque constant, and $\theta_l$ is its angular rotation at the load.

More complex models may be constructed (Electrocraft, 1980) to account for frictional and other losses. However, for most applications, these extra effects may be neglected, as may be the inductance of the armature and the elasticity of the output shaft. Letting $J = J_l + J_m$, the following two equations then apply:

$$Ri = v - K_g\dot{\theta}_m$$

and

$$J\dot{\theta}_{\mathrm{m}} = K_{\mathrm{m}} i$$

These equations are represented in the block diagram of figure 3.6 and will be referred to in chapter 4 when considering the dynamic behaviour of a complete robot. Continuing the simplification process by eliminating $i$ yields

$$J\ddot{\theta}_{\mathrm{m}} + K_{\mathrm{g}} K_{\mathrm{m}} \dot{\theta}/R = K_{\mathrm{m}} v/R$$

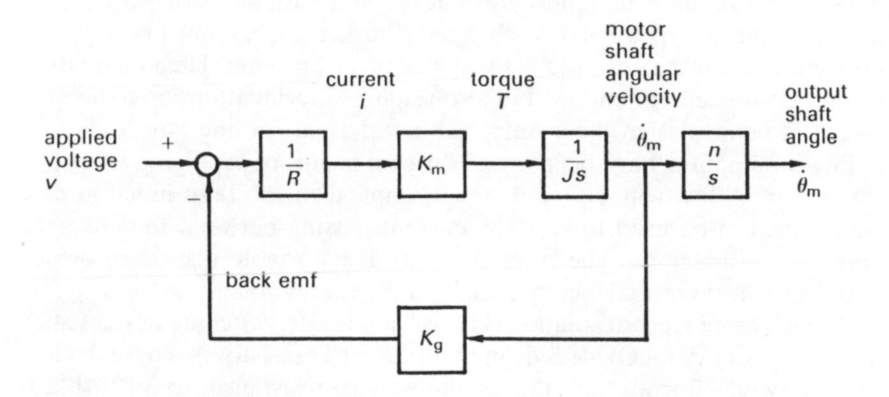


$K_{\mathrm{m}}$ = motor constant
$K_{\mathrm{g}}$ = generator constant
$R$ = armature resistance
$J$ = effective inertia of motor and load as seen by motor
$n$ = gear ratio

Figure 3.6 Block diagram for a DC motor

Taking Laplace transforms, with zero initial conditions, results in the transfer function

$$\frac{\theta_{\mathrm{m}}(s)}{V(s)} = \frac{K}{s(1 + s\tau)}$$

where the *motor time constant,* $\tau = JR/K_{\mathrm{g}}K_{\mathrm{m}}$ and $K = 1/K_{\mathrm{g}}$. Note that there are $i^2R$ energy losses in the armature. Limits will normally be put on the current $i$ to avoid overheating.

### *3.3.2 Other types of DC motor*

Another type of DC motor used in robots is the 'pancake type' (Margrain, 1983) which gives high torques at low speeds, and therefore may be used without a gearbox. The 'brushless DC motor' (Horner and Lacey, 1982) uses permanent magnet material in the rotor. The stator magnetic field is electrically produced, using coils which are energised such that the field rotates. No commutator or brushes are then necessary.

### *3.3.3 Power amplifiers and switching circuits*

The DC motor transduces electrical energy into mechanical energy. Some means must be used to supply this energy in a variable form so that the motor torque and speed may be changed. Furthermore, it must be supplied efficiently otherwise power losses may become a problem. Heat dissipation is not only wasteful of energy, but also requires special attention to be paid to circuit board design, heat sinks, cabinet design, cooling fans, etc.

Power amplifiers have been the traditional means of supplying energy to the motor. A low-voltage input signal, applied to the base junction of a transistor, can be used to vary the current passing between the collector and emitter junctions. The transistor acts as a variable resistance device and, as such, there may be appreciable $i^2R$, energy losses.

A much more elegant solution is to use solid state switching devices such as MOSFETs (Metal Oxide Silicon Field Effect Transistors). These devices may be switched from 'on', that is almost zero resistance, to 'off', that is very high resistance, very quickly. As the transistion time is very small and the 'on' resistance is very low, losses are kept small. If the transistor is switched rapidly between the on and off states, the average output voltage will be the ratio of the on to off times in the cycle, multiplied by the on-voltage. The switching signal input to the transistor is of low voltage, and it is easy to generate a sequence of pulses with the correct on to off ratio, in response to a demand voltage from the controller. Movement in the reverse direction is achieved by duplicating the circuitry such that the armature current is in the opposite direction in the second circuit. Care must be taken that only one of the circuits is operative at any given time.

## 3.4 Stepping motors

The main feature of a stepping motor is that, given a sequence of pulses, the output shaft will rotate by an integer multiple of the step size, which is typically between 1.5 and 30 degrees. The rotor comprises a permanent magnet with its magnetic axis lying along the axis of rotation of the rotor. Each end of the magnet is shaped into a toothed wheel, with the North-pole teeth radially displaced one-half tooth pitch from the South-pole teeth. Seen from one end, the rotor gives the appearance of a set of North–South poles distributed around the periphery, as seen in

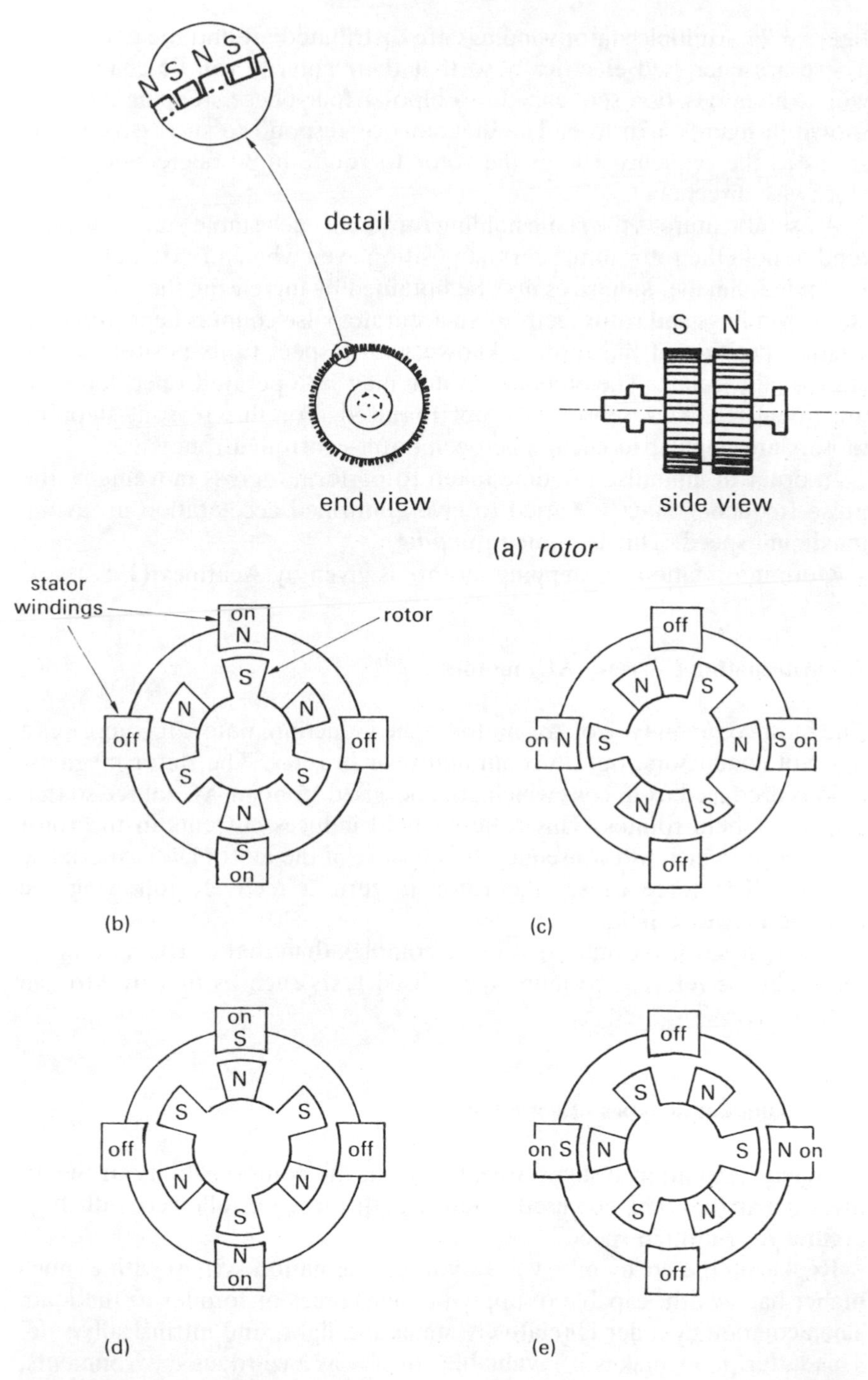


Figure 3.7 Stepping motor; (a) rotor; (b)–(e) energisation sequence

figure 3.7a. Multiple stator windings are distributed around the periphery. These are energised electrically so that their polarity can be changed at will. An energisation sequence for a bipolar four-phase stepping motor is shown in figures 3.7b to e. The diagrams correspond to successive stable states as the sequence causes the rotor to rotate in 30 degree steps in a clockwise direction.

A useful feature is the static holding torque in each stable state. This will tend to hold the rotor in the correct position even when an external torque is applied. Smaller step sizes may be obtained by increasing the number of stator windings and rotor teeth. If an accurate pulse count is kept, then the angular position of the rotor is known with respect to its position at the start of the count. The stepping motor may be operated open loop, so positional feedback sensors are not required. For this reason, stepping motors are popular in cheap microcomputer-controlled robots.

In order to minimise the time taken to perform a gross movement, the pulse frequency may be varied to give controlled acceleration up to the maximum speed. This is termed *ramping*.

More information on stepping motors is given by Acarnley (1982).

## 3.5 Alternating Current (AC) motors

The most common type of AC motor is the induction motor. It comprises a rotor of conductors, but no commutator or brushes. The stator magnetic field is produced from coils which are energised from an AC source so that the stator field rotates. This rotating field induces currents in the rotor conductors, which will consequently (because of the motor law) experience a force. The force causes the rotor to turn, effectively following the rotating magnetic field.

Full analysis of AC motors is more complex than that of DC motors, so the reader is referred to more specialised texts such as that by Morgan (1979).

## 3.6 Evaluation of types of actuator

Pneumatic actuation systems are cheap, but the compressibility of the air means that they are not used when a path must be followed with high accuracy, or at high speed.

Replacing the air by oil gives a hydraulic actuation system with a much higher bandwidth, capable of applying high forces or torques to the load. The actuation cylinder is relatively small and light, and intrinsically safe. This latter point makes it invaluable for use in hazardous environments, particularly if explosions could be set off by an electrical discharge.

However, a hydraulic power source must usually be purchased and a strict maintenance schedule adhered to in order to ensure reliability.

The DC motor is clean and can be used to drive a load with high precision. High torques are available either by using specialised types of motors or by using a gearbox.

AC motors are becoming more commonly used. No commutators or slip rings are required for induction motors.

All the above types of actuator may be used in closed loop control systems. A sensor is used to detect the output position, and its signal is used to drive the input in an appropriate manner until the output reaches the desired position. Stepping motors do not require such feedback. They can be used in open loop operation provided that an accurate count can be kept of the number of steps taken. However, the motor must not be stalled, since the pulse count would not then correspond to the actual number of steps rotated. In addition, for a given torque, a stepping motor is generally larger and heavier than a corresponding DC motor. By their very nature, stepping motors have a jerky action.

## 3.7 Transmission systems

If an actuator is to be connected directly to a joint, it must be able to provide sufficient force or torque, especially at low speeds. If this is not the case, then a gearbox may be used to gear down the output, thereby reducing the speed and increasing the force or torque seen by the link. Motion conversion may be necessary, so that a linear actuator may be used to produce a rotary motion, or a rotary actuator to drive a prismatic joint. It is desirable that all masses, of which the actuators will be a significant part, be located as close to the base as possible in order to reduce inertias. The movement from the output of the actuator must then be transmitted by some means to the remote joint.

### *3.7.1 Gearboxes*

Three types of gearing are commonly used: spur gears, epicyclic gears and the 'harmonic drive'.

*Spur gears* are shown in figure 3.8. If gear A has $m$ teeth and gear B has $n$ teeth round the circumference, then gear B will rotate at $-m/n$ times the rate of rotation of gear A. The gear diameters are proportional to the number of teeth, so if a torque $T$ is applied to gear A in a clockwise direction, this will result in a downwards force $T/r_A$ being applied to the edge of gear B at the point of contact. This will then cause an anticlockwise torque $Tr_B/r_A = Tn/m$ to be applied to gear B. In robotics it is usual to

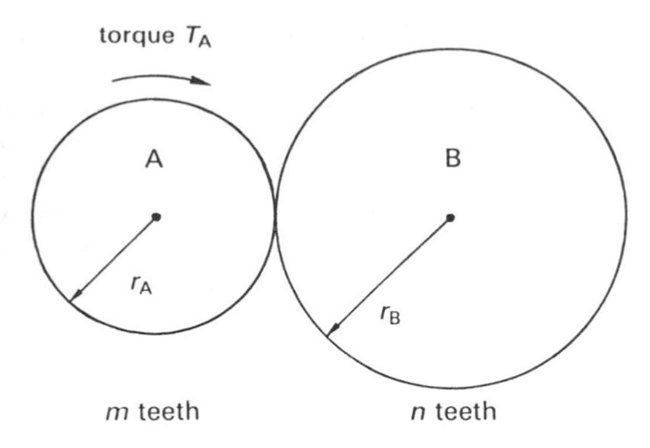


Figure 3.8 Spur gears (teeth round peripheries not shown)

have $m < n$, where gear A is on the actuator side and gear B is on the load side. Several pairs of spur gears may be combined in series to give a *gear train*.

If an inertial load $J_B$ is attached to the shaft of gear B which has an angular rotation of $\theta_B$, then

$$\ddot{T}_B = J_B\ddot{\theta}_B$$

where $T_B$ is the torque applied to gear B by gear A, as described above. A torque $T_A$ is applied to gear A by the motor. The shaft of gear A, with angular rotation $\theta_A$, will accelerate at $\ddot{\theta}_A$, and gear A will then have an apparent inertia $J_A$, given by

$$J_A = T_A/\ddot{\theta}_A$$

From earlier discussions, $T_A = T_B m/n$ and $\theta_A = \theta_B n/m$. Substitution in the above expression for $J_A$ gives

$$J_A = -(m/n)^2 J_B$$

The load inertia, as seen by the motor, may thereby be drastically reduced by appropriate gearing down.

In practice, the gears will not mesh together perfectly. Whenever movement of the input shaft is reversed, the slackness between the gears must be taken up by the input shaft before the output shaft reverses. This gives rise to *backlash*.

If the forces between the gears are low, for example when the load is a rotary position sensor, then *anti-backlash* gears can be employed. These use double gears to replace the single gear A. The double gears are sprung apart so that contact is always made on both sides of the teeth of gear B. Thus no slackness occurs, provided that the spring force always exceeds the applied force. The price to be paid lies in extra wear to the gears. This wear is related to the forces between the gears, and is usually excessive unless low driving forces are required.

*Epicyclic gears* are shown schematically in figure 3.9. They have an advantage over spur gears of co-axial input and output shafts. The usual way of obtaining speed reduction is to fix the annulus, attach the sun to the input shaft, and connect the output to the carrier which holds the freely rotating planet gears. The speed reduction will then be $((A/S) + 1){:}1$, where $A$ is the number of teeth on the annulus and $S$ is the number of teeth on the sun. Again, there will be some backlash.

*Harmonic drive* gearboxes are becoming popular in high-performance robots, the reason being their low backlash. The principle of operation is illustrated in figure 3.10. The harmonic drive comprises three parts. The first is a fixed annulus called the *circular spline*. Meshed into the annulus teeth is the *flexispline*, a non-rigid gear having two fewer teeth than the circular spline. The flexispline is pressed against the circular spline by the *wave generator* which is rigid and elliptical and acts as the input shaft. The output shaft, co-axial with the input, is connected to the flexispline.

As the wave generator is rotated through 360 degrees with respect to the circular spline which has, say, 100 teeth, these teeth must be engaged by 100 teeth on the flexispline. However, the flexispline has only 98 teeth so it must rotate the equivalent of two teeth relative to the fixed circular spline. A full revolution of the flexispline around the 100 teeth of the circular spline will occur after 50 revolutions of the wave generator. The harmonic

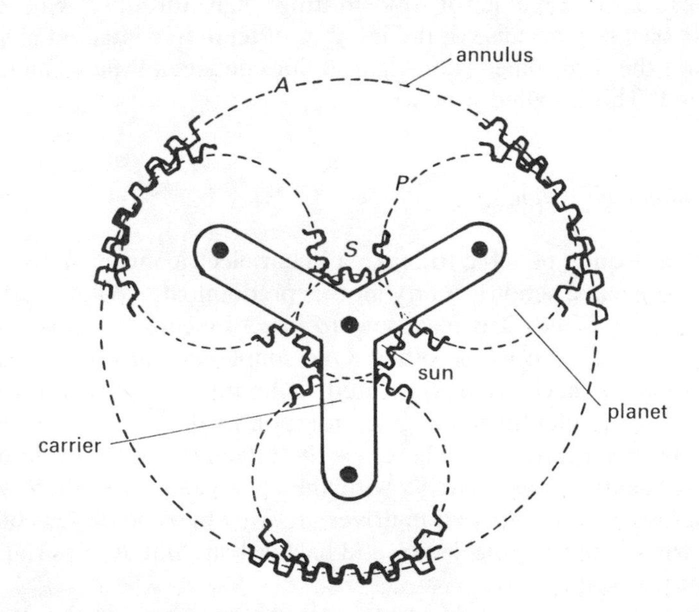


Figure 3.9 Epicyclic gears

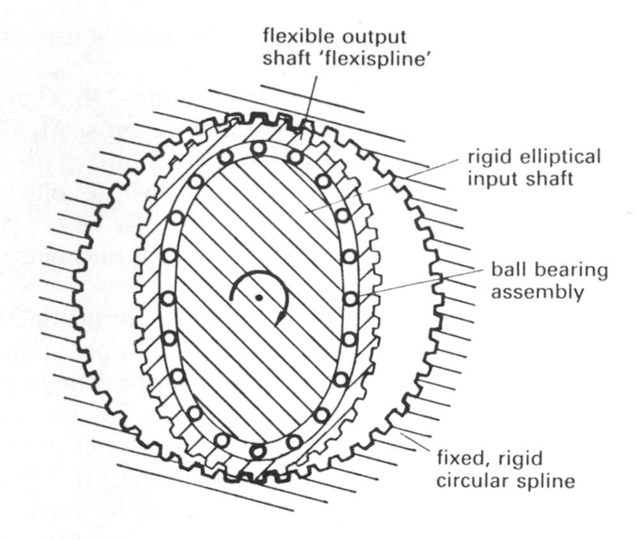


Figure 3.10 Harmonic drive gearbox

drive therefore provides a very compact means of obtaining large gear reductions. It is capable of transmitting high torques, with almost negligible backlash because of the number of teeth in contact at any given time. Since the flexispline is non-rigid, it does deflect a little when torques are applied. This is called *windup*.

### *3.7.2 Motion conversion*

Although it is quite possible to make a linear electric motor (linear in the sense of a prismatic action), rotary motors predominate. When a prismatic movement is required, it is necessary to have a means of converting the motion from one form to the other. One simple way of doing this is the *rack and pinion* mechanism illustrated in figure 3.11. The ratio of the movement of the rack to the angular rotation of the pinion is given by $r$ units of distance per radian, where $r$ is the effective radius of the pinion. Note that backlash will occur as with the spur gears, and there will be frictional losses. Rack and pinion drives are used in some designs of robot gripper, where the frictional losses and backlash are not as important as in the major axis drives.

A worm gear may be used in place of the spur gear in the rack and pinion. This gives a lower ratio of rack to shaft motion. The friction is high

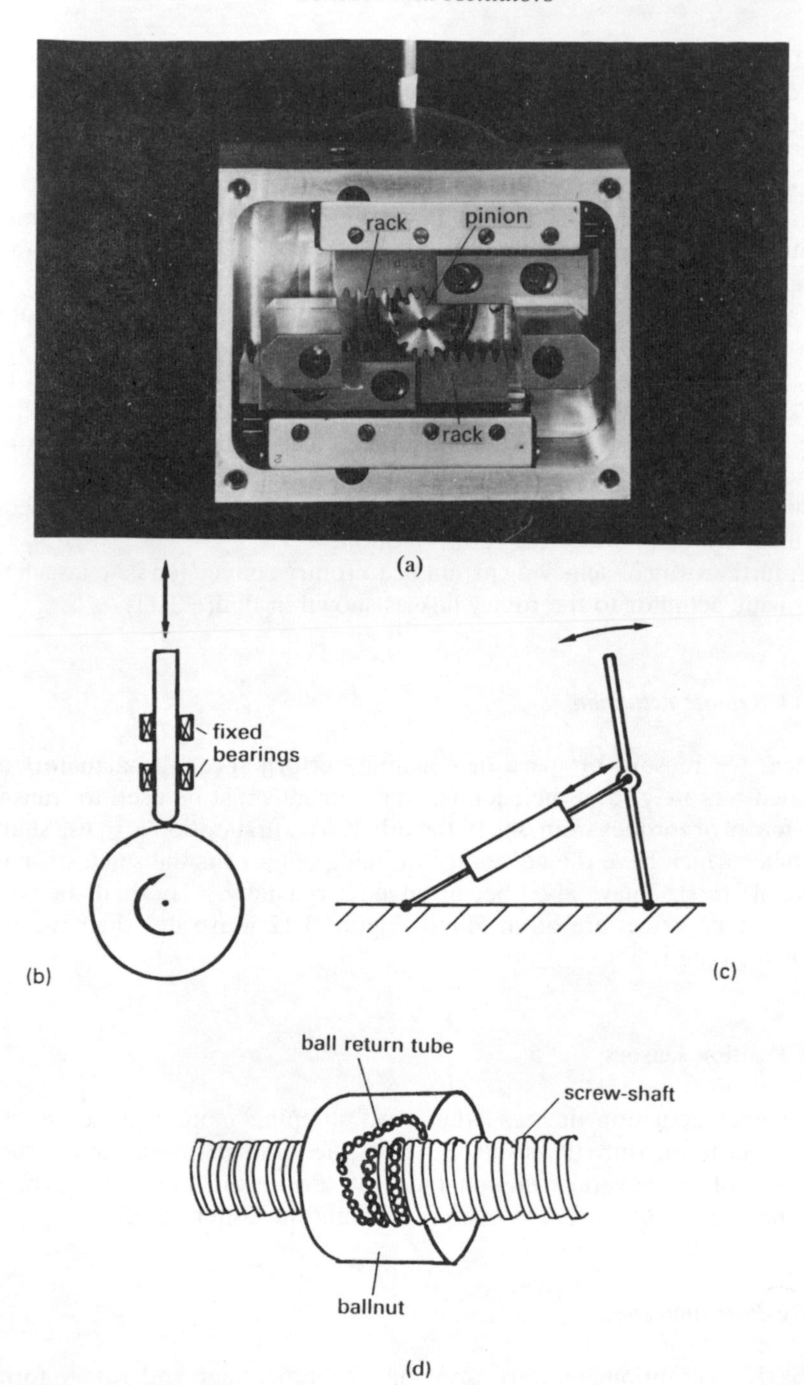


Figure 3.11 Motion conversion: (a) rack and pinion; (b) cam; (c) prismatic to rotary conversion; (d) ball screw

and may be used to good effect, in that the worm gear cannot be driven in reverse by the load. Thus, no holding torques need to be applied to the shaft in order to keep the rack in position.

*Cams* are a high friction solution to the problem, and are only seen in grippers. The principle of operation is shown in figure 3.11.

*Ball screws* offer a low-friction means of converting rotary to prismatic motion. The balls rotate round the screw thread as it is being turned, and a return path is provided to bring them back to the start of the bearing.

In contrast to the electrical actuator, hydraulic and pneumatic actuators are most commonly seen in their linear motion forms. Note that the rack and pinion mechanism, described previously, may be back-driven by the rack to cause the pinion to rotate. The rack may be replaced by a chain, or the rack and pinion by a wire and pulley. The principle of operation is exactly the same, but as the chain and wire cannot be compressed, the actuator must work against a spring, or in conjunction with a second actuator which provides the return motion.

A further way of achieving prismatic to rotary conversion is to attach the prismatic actuator to the rotary link as shown in figure 3.11.

### *3.7.3 Remote actuation*

When, for reasons of space or dynamic performance, the actuators are located remotely from their joints, some means must be used to transmit the forces or torques from one to the other. The first option is to use shafts, or tubes which have the advantage of being lighter for the same stiffness. Several tubes may also be arranged co-axially. Toothed belts or pulley-cable drives are often found. Figure 3.12 illustrates their use in a SCARA type robot.

## 3.8 Position sensors

Whenever actuation devices other than stepping motors are used, it is necessary to measure the displacement of the joints in order that the robot be controlled. Several common types of sensor will now be described. Further details of such sensors may be found in Usher (1985).

### *3.8.1 Potentiometers*

Resistive potentiometers are available in both linear and rotary forms. Considering figure 3.13, if the wiper is connected to a circuit having a very

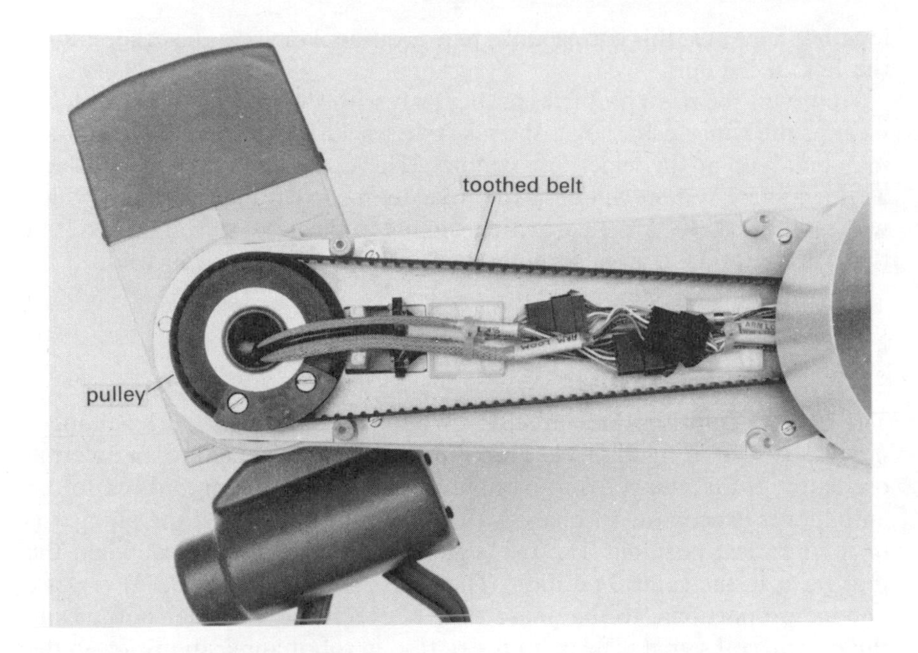


Figure 3.12 Toothed belts

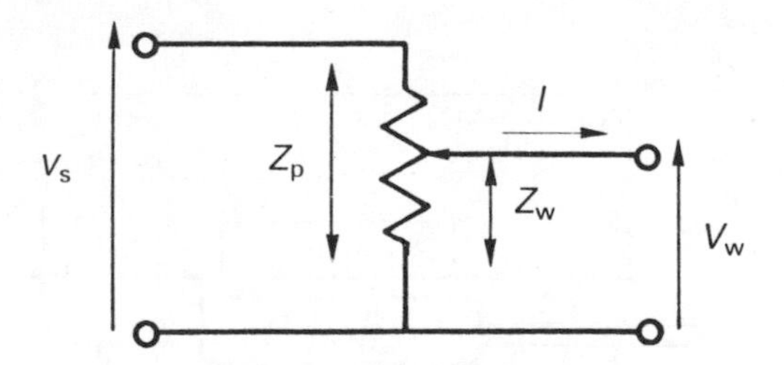


Figure 3.13 Potentiometer equivalent circuit

high input impedance, such as an operational amplifier, the wiper voltage is given by

$$V_w = V_s Z_w / Z_p$$

Note that this is only valid when $I = 0$. If $Z_w$ does not vary linearly along the resistive track, then the measurements will be in error. A potentiometer with 0.1 per cent linearity may seem an excellent device, but if the range of the potentiometer is 300 degrees this corresponds to a possibility of a 0.3 degree error. If the rotary joint is at one end of a link of length 0.5 metres, this corresponds to a positional error at the other end of the link of 2.6 mm.

However, the main problem, particularly with cheap devices, arises from wear as the wiper passes over the resistive track. In severe cases, the track may break up or the wiper lose contact. This can have a disastrous effect, as the control system will be given a vastly incorrect measurement which will cause it to give large correcting signals to the actuators. The result of this can be to make an apparently benign robot suddenly run wild.

### 3.8.2 *Linear Variable Differential Transformers (LVDTs)*

The LVDT comprises three precision-wound coils through which a plunger moves, as seen in figure 3.14. The central coil is connected to an external oscillator. If the plunger is at its rightmost position, the output fed to the amplifier is exactly out of phase with the output obtained if the plunger is in the leftmost position. The two signals should exactly cancel when the plunger is in the central position. The phase shift detector (PSD) yields a voltage proportional to the phase shift between the amplifier output and the central coil signal. The main restriction in robot applications lies in the available lengths, typically 1 mm to 20 cm, of commercially produced devices.

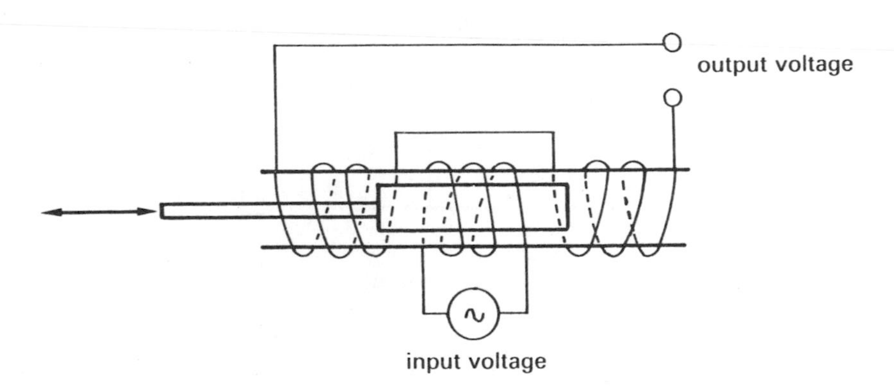


Figure 3.14 LVDT

### 3.8.3 *Optical encoders*

Optical encoders are available in incremental and absolute forms. Linear and rotary types may also be obtained. They have the advantage of being contactless, so avoiding wear problems, but must be kept clean, otherwise a faulty position will be detected.

The *incremental encoder* is the simplest device. In its rotary form, slots are made in the periphery of a disc as shown in figure 3.15. A photocell may then be used to count the slots as the disc is turned. It is possible to detect the direction of rotation by adding an extra photcell. A simple counting circuit may be used to find the angular displacement relative to the start-of-count position.

The *absolute encoder* enables the position to be detected at any time. Multiple tracks are usually provided as illustrated in figure 3.16. Historically, the Gray code has been used so that ambiguities are avoided when the photocells move from one sector of the disc to another. It is common, nowadays, for the output of an optical encoder to be read directly by a microprocessor.

Figure 3.15 Rotary incremental encoders

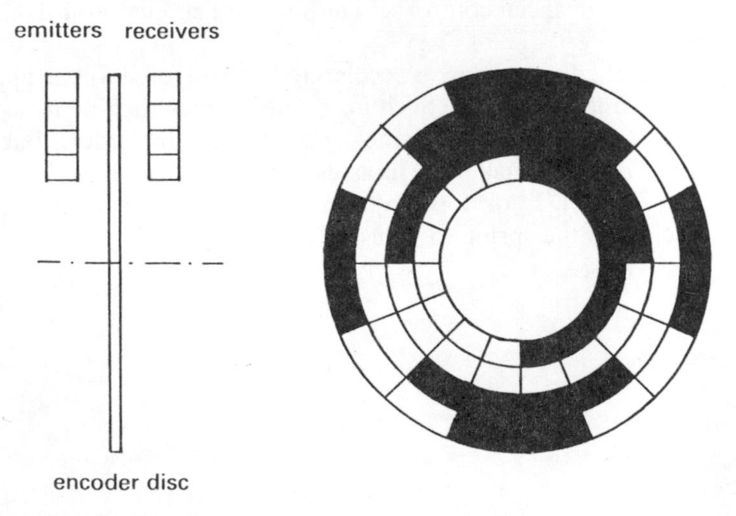


Figure 3.16 Absolute encoder

### 3.9 Velocity sensors

It will be seen in chapter 4 that velocity feedback is a feature of many robot axis controllers. If a permanent magnet DC motor is utilised, then a coarse measure of its shaft velocity may be obtained from a measurement of the back e.m.f. Alternatively, the difference between successive samples of position may be used to compute the velocity. Discretisation problems may arise if this technique is used when digital signals are taken, for example, from an optical encoder. On the other hand, a position sensor such as a potentiometer produces an analogue voltage output on to which electrical noise will be superimposed. The differential of this combined signal may be markedly different from the desired differential of the underlying signal, and thus large inaccuracies can occur when noise is present. The alternative is to employ purpose-made devices for velocity measurement. Rotary devices are the most common and they can conveniently be attached to a motor shaft. A linear to rotary motion conversion mechanism, such as a rack and pinion, would be utilised for prismatic joints.

A DC tachometer is essentially a permanent magnet DC generator. The generator law, of equation (3.7), states than an e.m.f. will be produced across the armature, and that it will be proportional to the speed of rotation. Unfortunately, a tachometer adds to the cost, bulk and weight of the actuation system.

### 3.10 Accelerometers

Measurements of acceleration are not used in most of today's robots but several proposed advanced control schemes require acceleration signals to be available.

If a mass, attached to a spring, is accelerated, then a force will be applied to the spring and it will extend. Such an extension may then be measured and should be proportional to the acceleration. In order that the accelerometer remain small, low masses are used. The mass may be connected to an LVDT, or a piezo-crystal. Alternatively, strain gauges may be attached to the spring to measure its extension.

# 4 Dynamic Control

A typical robot task comprises a sequence of movements. The purpose of this chapter is to describe how each movement is achieved through control of the individual joint actuators. Methods of commanding movement sequences are discussed in chapter 5.

Given a particular demanded movement, the robot controller must send appropriate signals to the joint actuators. For all but the most simple robots, the controller will also take signals from internal sensors, giving the feedback structure shown in figure 4.1. The design of the controller should be such that the robot meets the various performance criteria set out in chapter 1. The controller will usually be implemented on one or more digital computers.

Before considering the dynamic control of a robot, some of the essential features of linear feedback control will be reviewed.

### 4.1 Review of linear feedback control theory and techniques

The control of a single axis is considered here, using linear models of the actuator, load and sensors. The purpose of the controller is to enhance the performance in terms of stability, steady state errors, sensitivity to variations in the actuator/load/sensor parameters, sensitivity to external disturbances, and the speed of the transient response.

Figure 4.2 shows a block diagram of a simple feedback control system. Analysis of this figure allows an expression for $H(s)$, the *closed loop transfer function* to be derived as

$$H(s) = \frac{K(s)G(s)}{1 + K(s)G(s)F(s)} \qquad (4.1)$$

If the denominator is a second-order polynomial and the numerator is of zero order, this equation may be written in a standard form:

$$H(s) = \frac{K}{s^2 + 2\left(\dfrac{\zeta}{\omega_n}\right)s + \omega_n^2}$$

where $\zeta$ is the *damping ratio* and $\omega_n$ is the *undamped natural frequency*.

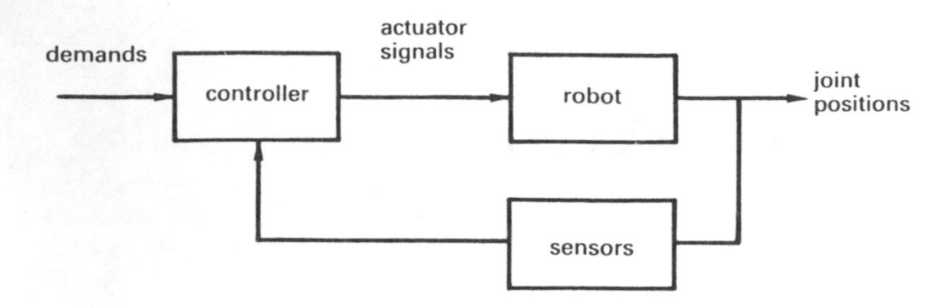


Figure 4.1 Simplified robot dynamic control

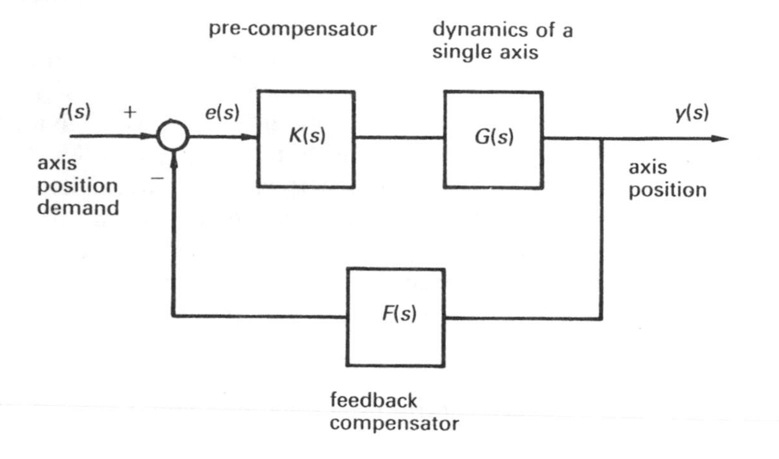


Figure 4.2 Simple feedback control system

The denominator of $H(s)$ is known as the *characteristic polynomial*. The location of its roots, the *closed loop poles*, determine the closed loop stability and the speed of the transient response. A detailed discussion of this, and the control concepts which follow, may be found in many control engineering textbooks such as that by Frankin *et al*. (1986).

### 4.1.1 Stability

Closed loop stability may also be determined from the frequency response data of the open loop system. When $K(s)G(s)F(s)$ is open loop stable, as is the case with industrial robots, the simplified Nyquist stability criterion may be used. This may be stated as follows:

'If $K(j\omega)G(j\omega)F(j\omega)$ is traced out from low to high frequencies in the complex plane, the closed loop system will be stable if and only if the *critical point* $(-1,j0)$ lies to the left of this locus.

Figure 4.3 shows a typical open loop frequency response plot. It is useful to know, not only if the closed loop system is going to be stable, but also to have some indicators of the closeness to instability. Two such measures of this, the *gain margin* and the *phase margin*, are commonly used, and are defined on the figure.

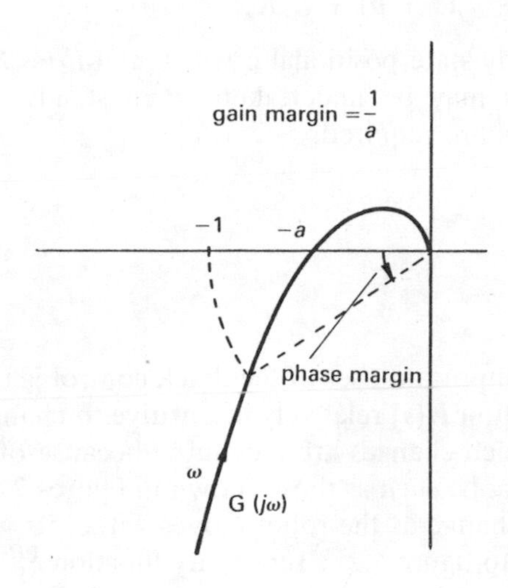


Figure 4.3 Open loop frequency response

### *4.1.2 Steady state errors*

In practice, a joint never comes *exactly* to rest at the commanded position. A *steady state error* arises because of frictional effects and measurement errors; it may also occur even if the ideal linear model is used, as will now be shown.

By examination of figure 4.2, the 'error signal' $e(s)$ may be derived as

$$e(s) = \frac{r(\mathrm{s})}{1 + K(s)G(s)F(s)}$$

Note that $e(s)$ is truly the error signal $(y(s)-r(s))$ *only* when there is unity feedback. The *final value theorem* lets us find $e(t)$ as $t \to \infty$.

$$\lim_{t \to \infty} e(t) = \lim_{s \to 0} se(s)$$

For example, if $r(s) = 1/s$, a unit step, and $F(s) = 1$, and

$$K(s)G(s) = \frac{K_p K_g}{(s + a)(s + b)}$$

then $e(s)$ is given by

$$e(s) = \frac{(s + a)(s + b)}{s\{(s + a)(s + b) + K_p K_g\}}$$

and then the steady state positional error = $ab/(ab + K_p K_g)$.

Similar analysis may be undertaken when steady state velocity and acceleration errors are required.

### 4.1.3 Sensitivity

One of the most important uses of feedback control is to make the closed loop transfer function $H(s)$ relatively insensitive to changes in $G(s)$. In the case of a robot, such changes are inevitable because of varying loads. In serial-topology robots, such as those shown in figures 2.7 and 2.8, the link inertias will also change as the robot moves within its working volume.

With reference to figure 4.2, a sensitivity function $S_G^H$, of $H$ with respect to $G$, may be defined as

$$S_G^H = \frac{\mathrm{d}H}{\mathrm{d}G} \times \frac{G}{H} = \frac{1}{1 + KGF}$$

This shows that $H$ will be insensitive to $G$, provided that the *loop gain* $KGF \gg 1$. Thus uncertainty in $G$, and similarly in $K$, may be tolerated if the loop gain is very high. Unfortunately, there will be limits to the allowable gain which can be used, either because of practical limits to signals into actuators, or because of stability constraints. Large loop gains are particularly difficult to achieve at high frequencies as, in practice, $G(\mathrm{j}\omega) \to 0$ as $\omega \to \infty$.

Similar analysis shows that

$$S_F^H = \frac{-KGF}{1 + KGF} \approx -1 \text{ if } KGF \gg 1$$

So, if the loop gain is high, any variation, say 0.1 per cent in $F$ will alter $H$ and therefore the output $y$ by $-0.1$ per cent. Thus, the joint position sensors of a robot must all be very accurate in order to achieve precise control of the end effector.

### 4.1.4 Forward path compensation

In order to improve the closed loop behaviour, a *compensator*, or *controller*, $K(s)$, may be inserted into the forward path as shown in figure 4.2. The '*PID*' (Proportional/Integral/Derivative) controller is considered here although several other types are commonly used.

The first subtype of the PID controller is the '*P*' (Proportional) type controller corresponding to $K(s) = K_p$, a pure gain. Choice of a large value of $K_p$ may be used to decrease the steady state error and effectively fools the plant into thinking that the error is larger than it actually is. This causes the output to respond faster to any sudden changes in the demand signal. However, large gains usually have a destabilising effect, and there may still be an unacceptable steady state error.

The '*PI*' (Proportional and Integral) type is used to remove any steady state errors. In this case, $K(s)$ is given by

$$K(s) = K_p \left(1 + \frac{1}{T_I s}\right)$$

The removal of the steady state error can be confirmed by considering the final value theorem, or by noting that the output can only reach a steady state when the actuator is at a steady state. When integral action is present, this can only occur when the error signal is zero. This type of control also has a destabilising effect.

The '*PD*' (Proportional and Derivative) type is used to increase the damping in the response via the derivative term, and thereby has a stabilising effect. The compensator has the form

$$K(s) = K_p(1 + T_D s)$$

The '*PID*' (Proportional, Integral and Derivative) type aims to combine the best properties of the PD and PI types. The integral action removes steady state error, and a combination of the P and D terms is used to give a fast, stable and satisfactorily damped response. The transfer function of a full PID controller is given by

$$K(s) = K_p \left(1 + \frac{1}{T_I s} + T_D s\right)$$

### 4.1.5 Other feedback compensation techniques

*Tachometer*, or *velocity* feedback, shown in figure 4.4, is commonly used in robot axis control. If the proportional compensator–actuator–load combination has a transfer function

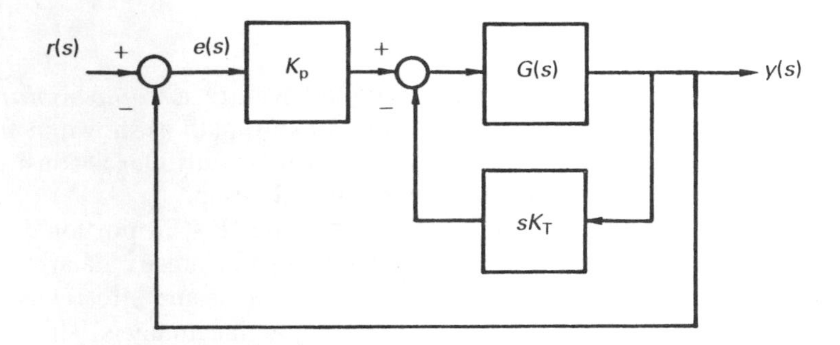


Figure 4.4 Velocity feedback with proportional control

$$G(s) = \frac{K_p K_g}{s(s + a)}$$

then, with unity feedback, the closed loop transfer function $H$(s) is given by

$$H(s) = \frac{K_p K_g}{s^2 + as + K_g K_p}$$

Alteration of the gain $K_p$ can therefore affect only one of the coefficients of the characteristic polynomial. If velocity feedback is added as shown in figure 4.4, then

$$H(s) = \frac{K_p K_g}{s^2 + (a + K_g K_T)s + K_g K_p}$$

Two of the denominator coefficients may now be changed, giving the designer more freedom to move the poles to speed up the transient response whilst retaining closed loop stability. A purpose-made velocity sensor of the type described in section 3.9 is usually attached to the joint,or joint motor, in order to obtain the required velocity signal.

Figure 4.5 shows both velocity and positional feedback applied to a single axis actuator–load combination. A permanent magnet DC motor, as described in section 3.3.1, is combined with a purely inertial load. The motor–load part of the block diagram is a representation of the motor equations (3.8).

Analysis of this block diagram shows that

$$\frac{q(s)}{r(s)} = \frac{nK_p K_m}{RJs^2 + K_m(K_g + K_m K_T)s + K_p K_m n} \tag{4.2}$$

The coefficient of the first-order term in the denominator is increased if viscous friction is not negligible. A full analysis of this case may be found in

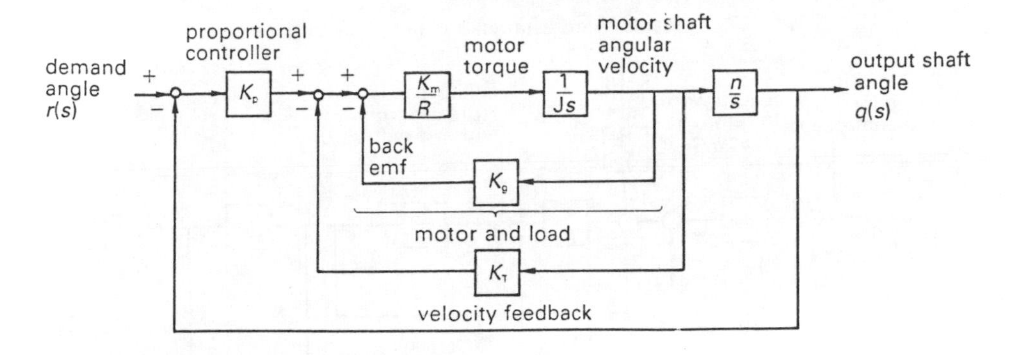


$K_m$ = motor constant
$K_g$ = generator constant
R = armature resistance
$J$ = effective inertia of motor and load (at motor)
$n$ = gear ratio
$K_p$ = proportional gain constant
$K_T$ = velocity feedback constant

Figure 4.5 Single axis control with position and velocity feedback

Luh (1983). A damping ratio greater than unity is specified so that no overshoot occurs. A rule of thumb (Paul, 1981) is to specify an undamped natural frequency of no more than half the structural resonant frequency.

A problem with robots is that the load carried by the end effector may be unknown. Also, the inertia $J$, as seen by the actuator, will vary according to the values of other joint values, One approach (Luh, 1983) is to use the maximum possible $J$, and the values of $K_p$ and $K_T$ are chosen to meet the above damping ratio and undamped natural frequency constraints.

A better approach has been used for the ASEA IRB 3000 robot. The weight of the end effector load is fed into the controller as part of the task description. The controller contains a simplified dynamic model of the robot and uses this to vary the values of the control coefficients as the robot moves from one part of the working volume to another. It has been reported (Bergman *et al.*, 1986) that savings of 20–40 per cent of travel times are possible using this method.

*Feedforward compensation* is often used when steady state errors and disturbances can be accurately predicted. The basic idea is to inject an extra signal into the existing closed loop, such that it nullifies the disturbance or steady state error. Figure 4.6 shows how this might be done for a single axis joint control subject to disturbances from other axes, friction and gravitational effects.

Clearly, if the estimates of the disturbances are incorrect, the nullifying signal which is injected will also be incorrect. The output will then still be subject to disturbances. If the estimates are grossly in error, then

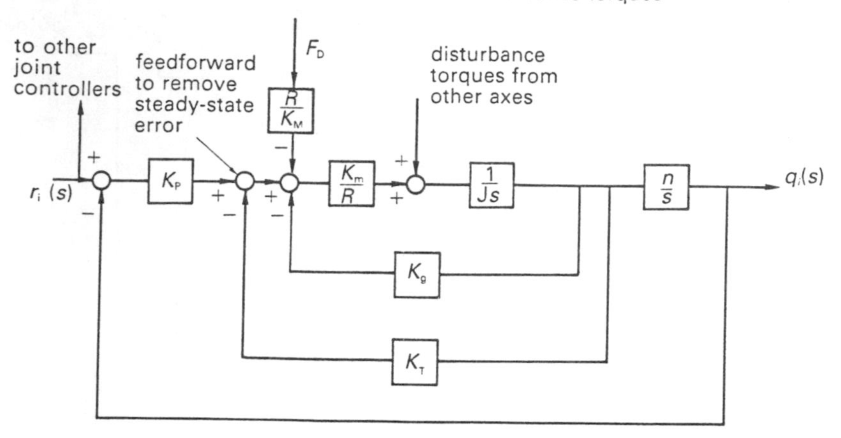


Figure 4.6 Feedforward added to the single axis control system

feedforward compensation can actually worsen the control. Particular difficulties arise when frictional components are compensated in this way.

Feedforward compensation may also be used to reduce the interaction between one robot axis and another. This is described in more detail in section 4.2.2.

### 4.1.6 Digital control

The above compensators may well be implemented on a digital computer. If the sensors produce analogue voltages then analogue-to-digital digital conversion is required. The output from the computer will be in digital form. This will require digital-to-analogue conversion unless, for example, the computer outputs a series of digital pulses to a stepper motor as discussed in section 3.3.

There is insufficient space here to discuss digital control in detail. Those wishing for more information should consult texts such as that by Houpis and Lamont (1985). This gives a particularly good account of the practicalities of software design and other implementation aspects.

An outline of the derivation of a digital version of a PID controlle: will suffice to illustrate some of the principles. The digital equivalent of the time derivative $e(t)$ of the error signal $e(s)$ is given by the following approximation (first-order backward difference):

$$\frac{e(k) - e(k-1)}{T} \tag{4.3}$$

where $e(k)$ is the current sample value of $e(t)$, and $e(k-1)$ denotes its previous sample value $T$ seconds earlier. If the output of the controller is given by $m(t)$, then for PID control

$$m(t) = K_p\left\{e(t) + \frac{1}{T_I}\int e(t)dt + T_D\dot{e}(t)\right\}$$

and therefore, on differentiating:

$$\dot{m}(t) = K_p\dot{e}(t) + \frac{K_p}{T_I}e(t) + K_pT_D\ddot{e}(t)$$

The terms for $\dot{e}$, $\ddot{e}$ and $\dot{m}$ may now be replaced by their digital equivalents in a similar manner to that shown in equation (4.3), giving the *difference equation*:

$$m(k) = m(k-1) + K_p\left(1 + \frac{T}{T_I} + \frac{T_D}{T}\right)e(k) - K_p\left(1 + \frac{2T_D}{T}\right)e(k-1) + K_pT_De(k-2) \tag{4.4}$$

The processor must store the two previous values of the error, namely $e(k-1)$ and $e(k-2)$, and the last value of its output, $m(k-1)$.

Note that the validity of this approach depends on the sampling frequency $1/T$ being much greater (typically ten times) than the frequencies present in the error signal.

## 4.2 Point-to-point control

It is now necessary to see how a required movement of the end effector may be achieved by means of control signals to each joint actuator. In *point-to-point mode*, the robot is commanded to move from one position to another without constraints on the path between the two. *Unco-ordinated control* describes the case when independent commands are input to each joint actuator. Typically, there would be a feedback control system for each joint. Consequently, each joint reaches its target position at a different time.

*Co-ordinated control* forces all joint motions to start together and to end simultaneously. The resulting action is much smoother than for unco-ordinated control. An algorithm for the co-ordinated control of a multi-axis robot driven by stepping motors is now described. This is followed by a description of multiple joint control, assuming that each joint cannot be satisfactorily controlled independently of the others.

### 4.2.1 Co-ordinated control of a robot driven by stepping motors

The most common way of generating co-ordinated motion is to first find the motor which must be driven the largest number of steps. Denote this number by the symbol $M$. A software loop (executed $M$ times) is generated whereby this motor is stepped once within the loop. Other motors are stepped less often, their steps being distributed evenly through the $M$ loop executions. This may be done as follows (Microbot Inc., 1982):

Set all motor counters to $M/2$

Repeat the following loop $M$ times
- Subtract the absolute value of the number of steps each motor is to be stepped from its corresponding counter
- If any motor counter is negative, then step that motor in the requested direction and add $M$ to its counter
- Enter a *Wait* routine to give a software time delay so that the motors can complete their movements before the next steps are commanded

### 4.2.2 Multiple joint control

The simple approach to the control of multiple joints assumes that all joints are independent of each other. That this is not necessarily so is particularly apparent in serial topology robots. In the example shown in figure 2.8 and analysed in section 2.6, there is significant interaction between the torques. A torque applied to joint $i$ of a robot will affect link $(i - 1)$ as well as link $i$. As the end effector is moved through the working volume, there will be significant changes in the load inertia as seen by the actuator at joint $i$. The conventional approach to this problem is well described by Luh (1983), and is outlined below.

The dynamic model for an $n$-joint manipulator is

$$F_i = \sum_{j=1,n} D_{ij}\ddot{q}_j + \sum_{j=1,n}\sum_{k=1,n} D_{ijk}\dot{q}_j\dot{q}_k + D_i \tag{4.5}$$

The $F_i$ are the forces or torques applied to the $i$th joints and $q_i$ is the $i$th joint position. The precise meanings of the $D$ matrices are given in section 2.6. In the single loop case, $F_i = D_{ii}\ddot{q}_i$, so most of the terms are actually neglected if independent single loop controllers are used. The remaining terms may be regarded as disturbance forces/torques to the single loop. In theory, their effect may be cancelled by injecting a feedforward signal $F_D$ as shown in figure 4.6. $F_D$ is given by:

$$F_D = \sum_{\substack{j=1,n \\ j \neq i}} D_{ij}\ddot{q}_j + \sum_{j=1,n}\sum_{k=1,n} D_{ijk}\dot{q}_j\dot{q}_k + D_i \tag{4.6}$$

The $D_{ij}\ddot{q}_j$ terms represent the coupling inertias. Given samples of $\dot{q}_j$, then $\ddot{q}_j$ may be deduced. The $D_i$ terms represent the gravitational terms which may be computed. It is conventional to neglect the centrifugal and Coriolis terms. The feedforward signal $F_D$ then becomes

$$F_D = D_i + \sum_{\substack{j=1,n \\ j \neq i}} D_{ij}\ddot{q}_j$$

The main difficulty in the above scheme lies in the use of the $D_{ij}$ terms. These are complex and their computation is time consuming. Luh (1983) describes various simplification techniques which have been used, or proposed, to make the computation possible within the sampling time.

As a result of all the simplifications made above, the control may be distinctively suboptimal, particularly at high speeds when the Coriolis forces become non-negligible.

The control problem becomes more difficult for the serial type robots, particularly the anthropomorphic type of figure 2.7. The couplings between control loops are less serious for SCARA types because of fewer interacting axes and the simpler dynamic model. Cartesian robots are particularly easy to control since there are no couplings at all between the first three orthogonal axes starting out from the base.

## 4.3 Continous path control

This type of control is used when a certain path must be followed between the endpoints. In welding, for example, the welding torch must follow the desired welding line at a prescribed speed. In assembly, it is common for straight-line motions to be specified, not only for insertion tasks, but also to avoid obstacles. In this case, intermediate *waypoints* may be defined through which the trajectory must travel. It is possible to control the end effector in world co-ordinates, or the path may be redefined in joint co-ordinates and the joints controlled directly.

### *4.3.1 Use of the Jacobian*

The first approach is to divide the Cartesian path (in world co-ordinates) into a number of segments. This gives a set of waypoints. If these waypoints are close enough together, and the end effector actually passes through them, then the complete desired path should be followed within the allowable error bounds. Each point is converted into a set of joint co-ordinates using the transformation techniques developed in chapter 2. If the segments are short, so that each $\delta q_i$ (the increment in the $i$th joint) is small, then the differential transformation, the Jacobian of section 2.5.5, is used. Since the $\delta q_i$ are small, the approximations $\sin \delta q_i \approx q_i$ and $\cos \delta q_i \approx 1$ may be made. Equation (2.5) repeated below, with $\delta \boldsymbol{q} = \delta \boldsymbol{\theta}$ and $\delta \boldsymbol{p} = [\delta x, \delta y, \delta z]^T$, is then linear.

$$\begin{bmatrix} \delta \boldsymbol{p} \\ \delta \boldsymbol{\theta} \end{bmatrix} = J \, \delta \boldsymbol{q}$$

In order to find $\delta q_i$ for a given $[\delta \boldsymbol{p}, \delta \boldsymbol{\theta}]^T$, the Jacobian matrix $J$ must be inverted. This gives rise to poor path control when close to a singularity, as discussed in section 2.5.5.

### *4.3.2 Use of curve fitting*

The approach described above requires either the storage of many pre-computed points, or their online computation. An alternative approach is to compute, offline, $n$ functions of approximation to the joint path, one function for each joint. These relatively simple functions are then evaluated online and used as the commands to the control system.

One method of such curve fitting is to split the path up into $n$ segments defined by the two end points and $(n - 2)$ waypoints. If the velocity and acceleration at the start and end of the path are zero, then the first and last polynomials are fourth order, and the remaining $(n - 3)$ polynomials are third order. This allows continuity of position, velocity and acceleration at waypoints (Edwall *et al.*, 1982). An excellent description of the derivation of the polynomial coefficients, together with a numerical example, is given by Ránky and Ho (1985).

# 5 Task Control

It will be assumed in this chapter that we have available a robot which can be commanded to move to certain positions or along certain paths. All the necessary hardware and software components, described in the previous chapters, have been assembled. For the moment we will assume that no external environmental sensing is necessary. However, some means must be provided for programming the robot to perform the required series of movements. This chapter is about elementary task control—how simple non-sensory tasks may be taught and executed. This does not necessarily preclude a small element of uncertainty in component locations. Compliance, whether arising from the robot structure, the workpiece or its jigs, or passive compliance devices as described in section 2.3.7, may all be used to assist the joining of parts during assembly tasks.

## 5.1 Teaching by guiding (teaching by showing)

This is the simplest form of teaching. The robot is guided manually through the set of points representing the task. The operator indicates to the robot controller when a point must be remembered. The set of remembered points, in the form of the joint positions, forms the program stored in the controller's memory. Execution of the program consists of the robot moving to each stored point in turn.

The guidance of the robot during teaching is often effected by means of a *teach box* or *teach pendant*. The Adept I teach box is shown in figure 5.1. This box is connected to the control cabinet via an umbilical cord, and it contains all the necessary functions to move the robot around the cell. Hand depression of the keys of the teach box causes signals to be sent to the controller which then activates the particular joint or set of joints. When the robot reaches a desired position, the location is stored by depressing another key.

In some robots, particularly those used for paint spraying, the operator moves the counterbalanced robot arm through the desired path. Co-ordinates along the path may be stored automatically.

Some robots allow movements to be commanded in one of several co-ordinate systems, typically joint mode and world mode, but sometimes additionally in tool mode. Joint mode allows individual joints to be moved.

World mode allows movements to be specified in Cartesian co-ordinates and usually roll, pitch and yaw angles. Tool mode allows movements to be undertaken with respect to a co-ordinate system attached to the end effector. This latter facility is especially useful in 'peg in hole' problems when the hole is not aligned with either the world or joint co-ordinate frames.

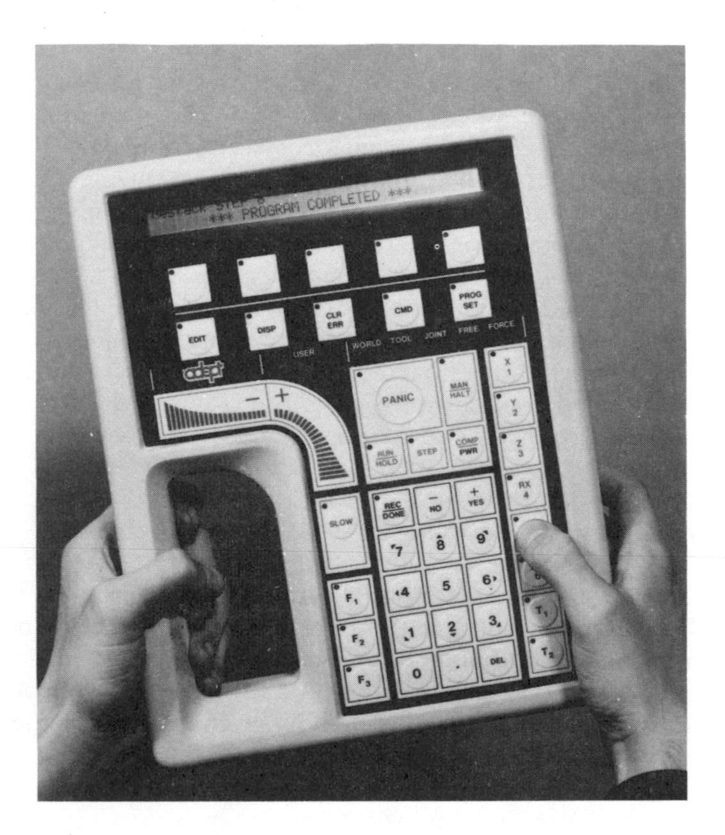


Figure 5.1 Adept I teachbox

In the majority of existing robots, teaching by guiding is the only way of specifying the task. In more advanced robots, the main task description is entered into the controller as a computer program. This is the subject of the next section. However, in this case various reference positions within the workcell may not be known in advance with the required precision. These positions may be treated within the program as variables. Their values can then be set through a 'teaching by guiding' initialisation phase.

## 5.2 Offline programming (manipulator level)

Teaching by guiding defines the task in terms of a series of geometrical points through which the end effector should pass. With offline programming, the task is defined by a computer program comprising a sequence of commands to be obeyed by the robot. Points may be defined as variables, allowing the geometrical information to be separated from the task description.

This section will describe *manipulator level programming*, so called because the task description is defined in terms of robot movements. The implications of external sensory feedback will be considered in chapter 6. *Object level programming* defines the task in terms of what must be done with the various objects in the workspace. *Objective level programming* describes the desired outcome of the task. These latter two levels are at a higher level than the manipulator level and are still the subject of research. They will be considered again in chapter 8.

The desirable properties of a language in which to write the task program, and the necessary supporting software, are now described.

### *5.2.1 Language requirements*

A good robot programming language must have all the data types and structural constructs necessary for the description and execution of the task. It should not have too many superfluous features. For example, it is confusing to the reader of a program if there are many commands which give almost the same result. The most important features of any computer program are firstly that the program should perform the function intended, and secondly that it is easy to understand and modify by someone other than the original programmer. Programs written with clarity and maintainability in mind are also more likely to work. Another factor which may be important is the efficiency of the code, although the speed of execution of the program is likely to be much faster than the mechanical speed of the robot. Transportability should become more important in the future. It would be desirable if a task description could be executed on different robots with the same result. However, this is probably best done at object and objective level programming. Since the kinematic structures and the working volumes of different robots can be very dissimilar, a manipulator-level program written for one robot may well be inappropriate for another.

There is no standard language used for manipulator-level programming. Some robot manufacturers provide their own language with the purchase of the robot, an example being the VAL language (Unimation Inc., 1985) provided with several of the Unimation robots, another being AML(Taylor, R. H. *et al.*, 1983) used with the IBM 7565 robot. Bonner and Shin (1982) give a comparative study of such languages. Another option is to use an existing general-purpose computer language such as Pascal or Forth as a basis for the manipulator-level language. A library of procedures is provided which provides the basic commands and interfacing with the robot control system. This approach is taken by Universal Machine Intelligence Ltd with their RTX robot.

There must, of course, be commands provided for motion control. The most common way of doing this is to provide a MOVE type command, with a set of arguments defining the position that the robot is to be moved to. It is useful to be able to specify this target position in one of several co-ordinate systems, world and joint modes being the most valuable. Both absolute and relative (with respect to the current position) movements may usually be defined. The program becomes more readable and easier to maintain if names are given to the positions. Those that are fixed can be defined at the start of the program. Those that are not can be treated as ordinary program variables. Integer and real data types are used for such variables. Character strings are used for message passing.

Branches and loops are useful even in programs which do not have to respond to sensory data. One example is a palletisation task where components are supplied from a fixed point and have to be placed into the pallet in one of an array of regularly spaced locations, as shown in figure 1.3. One method of programming this task is to set up two loops, the $x$ location being varied by one loop which is nested within the second loop which varies the $y$ location. Any of the usual DO–WHILE or REPEAT–UNTIL type structures may be involved for the loop control. The actual positions may be computed within the loop, thus needing support of arithmetic operations such as addition, subtraction, multiplication and division. Trigonometrical function support is helpful in cases such as that shown in figure 5.2. Here the pallet positions are most conveniently defined as a regular pattern in polar co-ordinates about the centre of the pallet. These co-ordinates must be transformed to a co-ordinate set in which the robot can be driven. If the set of pallet locations is irregular, and therefore not conveniently computable from a single starting point, an array of the locations relative to the starting point is probably the most convenient and versatile way of storing the positons.

Other program flow control statements which are useful are of the conditional or unconditional branching types such as IF–THEN–ELSE or GOTO forms. The use of branches implies some means of defining the point in the program to which control must pass. Line numbers may be used as in simple BASIC-like languages or, preferably, labels may be used.

The above features give the programmer great flexibility. The arguments of a MOVE command may be explicitly defined within the program, they

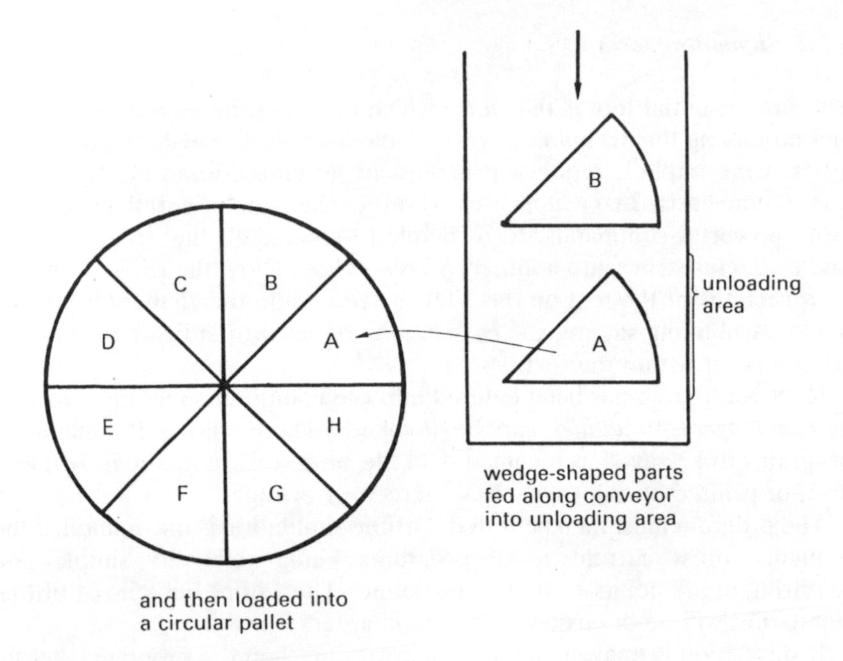


Figure 5.2 Palletising task

may be defined through a teaching phase, or derived mathematically from other variables, thereby minimising the number of taught positions. In addition, the programmer can combine the use of variables with loop and branch constructs, giving very versatile program flow control.

In order to improve program clarity and maintainability, the use of procedures and functions is beneficial. Furthermore, if these procedures can be collected together in libraries, then these libraries may be used by other programmers. The existence of fully debugged, well-documented and well-written libraries of procedures can greatly improve the productivity of the programmer and make his or her task so much easier.

Well-written programs contain many commenting lines, preferably about as many lines of comment as there are lines of executable code. Well-written programs are also well structured. This is mainly a function of good offline planning of the organisation of the program and the use of a consistent programming style aimed at making the program easy to understand, maintain and extend. It is well worth noting that the use of a 'structured' language does *not* automatically ensure a well-structured program in the above sense. It is quite possible to write a badly structured program with 'structured' constructs as it is to write a good program in an 'unstructured' language. Clear thinking by the programmer is the most significant need.

### 5.2.2 Supporting tools

The first essential tool is the *editor* which provides the means of entering and modifying the program. A very simple *line editor* may be used, which, as its name implies, requires programs to be entered and changed on a line-by-line basis. *Text editors* are text rather than line orientated and allow more powerful commands, for example to change all the occurrences of one character string into another. *Screen editors* allow the programmer to view sections of the text on the VDU screen while the changes are being made. A blinking square, the *cursor*, may be used to indicate the current editing point within the text.

Once a program has been entered into the computer's memory, various *task management utilities* can be invoked. These allow, for example, programs to be saved as a named data file on a storage medium such as a disc, or printed on the user's VDU screen or printer.

The program must then be tested. Offline simulation is uncommon at the moment, most current robot programs being relatively simple and requiring many points to be taught online. The future benefits of offline simulation will be discussed further in chapters 7 and 8.

If simulation is unavailable or inappropriate, then the program is tested 'live' on the robot. For safety, it is good practice to operate the robot at a low speed, holding the emergency stop button ready for any possible dangerous move. The robot, once powered up, should be regarded as being totally unpredictable.

Some means must be provided for 'executing' or 'running' the program. A utility is invoked which reads the program statements and converts them into internal commands which are sent to the robot controller to move the joint actuators. Internal memory locations are reserved for the particular variables used. This process is most commonly undertaken by an *interpreter*.

The *interpreter* utility reads in each line of the program in sequence. It checks that the syntax of the line is correct, and if not stops the interpretation with an appropriate error message to the operator. If the syntax is valid it causes execution of the statements in the line, be they robot movement commands, program flow control statements or arithmetic operations. The process of checking the syntax and converting the statements into internal forms can be slow when a large number of commands and qualifiers are available. This is particularly so if the command format is not rigidly constrained. At manipulator-level programming, this is not usually important as the speed of the mechanical movement of the robot is often the main constraint on task execution time. An alternative approach is to *compile*, offline, the task program into internal code which can then be executed more quickly. Syntax errors may also be detected at this stage. Various schemes are also possible which combine the two approaches.

*Debug aids* may be used to help find faults in the program. There are three types of fault. Poor programming or typing may result in a program line having an invalid syntax. This should be found by the interpreter or compiler as above. However, a typing or programming error may result, for example, in an incorrect name being given for a variable. If the name itself is valid, then the interpreter or compiler will find no error. The error will only be found (perhaps) when this piece of code is executed. This type of error can be very difficult to find and rectify, especially when the particular section of code is executed infrequently. It should be emphasised that *program validation*, that is checking that the program correctly describes the task as required by the programmer, should be rigorous in order to keep such bugs to a minimum.

A third type of error is that of a *task specification error*. These are particularly common when sensory robots are used, since the programmer may not have thought of all of the problems which may occur during task execution. Although the program itself will be correct, it will not meet the overall task objectives. This error will also be found during task execution.

In order to help the user debug a program at runtime some robot manufacturers provide a *single-step* capability. Each program step is executed only after a prompt is given by the user. In large programs, it is helpful if this action can be switched on and off within the program. Some manufacturers provide *trace* facilities whereby, after an interruption or a fatal error, the user is told at which program line the fault occurred. If the fault occurred within a procedure, then the traceback facility will give the name of the calling procedure with the line number at which the call was made. This can be repeated for each level of procedure call back to the main program at the top level.

Good *error reporting* is essential. The user needs to be told as clearly and precisely as possible the nature of the error detected, whether it be a simple syntax error in the program line, or that the robot is being commanded to move to a position outside its working volume. It should not be necessary to have recourse to a programming manual in order to find the meaning of a particular error code number.

Finally, the need for good documentation extends to the manuals which will have been provided with the robot. They, and any online documentation and help facilities which may be available, should provide an easy-to-understand introductory guide, a reference manual for the experienced user, and illustrative worked examples. These comments apply not only to the programming manuals but also to the hardware documentation.

# 6 External Sensors

The essential feature of robotic systems is their flexibility, allowing them to be easily reprogrammed to perform different tasks. It was seen in the previous chapter how variables can be used to define the positions to which movements are made. The provision of conditional branching also allows different paths to be taken through the program, dependent on the values of certain variables.

Let us now suppose that we cannot say in advance precisely where the component is to be picked up from in the example of section 5.2. This is not uncommon, as precise feeding of components can be expensive. Let us suppose that we can provide some means of sensing the location of the component, and its co-ordinates can be passed to the robot into a variable. We may then use this variable as the component-pick location, and the robot can cope with components presented imprecisely.

Another type of sensory feedback is used to monitor the correct operation of part of the task. An example here is to sense the forces on the gripping fingers as a peg is inserted into a hole. Excessive forces indicate jamming or the absence of a large enough hole. If this is detected, then appropriate error recovery action must be taken. At the simplest level, this comprises halting the program with an error message on the user's VDU. At a higher level, error recovery procedures try to continue the task correctly without the need for operator action.

The third aspect of sensing falls into the category of *inspection*. It may be necessary within the robotic workcell to check that components are free from certain faults before, after, or during manipulation. Such faults may cause problems in executing the robotic task, or they may affect the quality of the final product.

In summary, information from sensors can allow the robotic system to perform its required task even when there are uncertainties in its environment. This chapter describes some of the sensors used for this work and the extraction of the required information from the sensory data. The implications for task control are discussed in chapter 7 where an illustrative example is given.

## 6.1 Vision sensors

Vision is a very powerful means of sensing the environment. One of the main reasons for this is the development, which has taken place over many years, of high-quality vision sensors for the television industry. Video cameras are relatively cheap compared with the cost of a robot, and are readily available. The emergence of solid state sensors, again with the television/home video market in mind, has meant that compact and rugged vision sensing devices can be used for robotic applications. Vision is also valuable because it requires no contact between the sensor and the object, and data acquisition is fast. In video cameras, an image may be acquired every 20 ms.

Several types of vision sensor are available and will now be described. This is followed by a section on lighting, arguably the most crucial part of vision systems.

### *6.1.1 Very-low-resolution vision sensors*

The simplest sensing solution is usually the best, and indeed many visual sensing problems can be tackled satisfactorily by means of a single emitter and receiver pair. The schematic of figure 6.1 illustrates the *thru-scan* technique. Here, the passage of light from the emitter to the receiver is blocked when an opaque object passes between them, giving a binary decision.

The *retroreflective* technique shown in figure 6.2 generally uses an emitter and detector housed in the same case to detect the presence of a

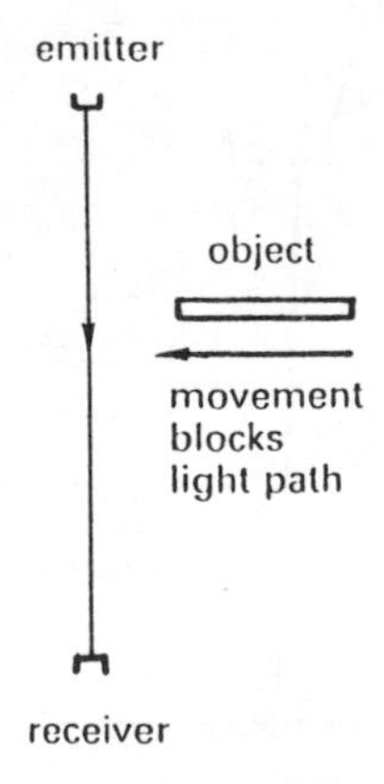


Figure 6.1 Thru-scan technique

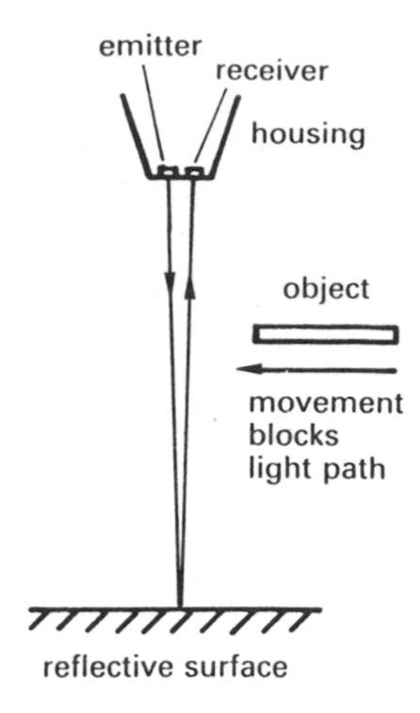


Figure 6.2 Retro-reflective technique

reflective surface. Again, if an opaque object is moved between the case and the surface then the light path will be broken.

The *diffuse reflective* technique, depicted in figure 6.3, senses light reflected from the surface of the object which should therefore be in front of a non-reflective background. If the reflective properties of the object are known, then the sensor may be used as a proximity sensor since the reflected light will reduce in magnitude as the object moves away.

The above devices usually operate in the infra-red region, thereby reducing the effects of background illumination.

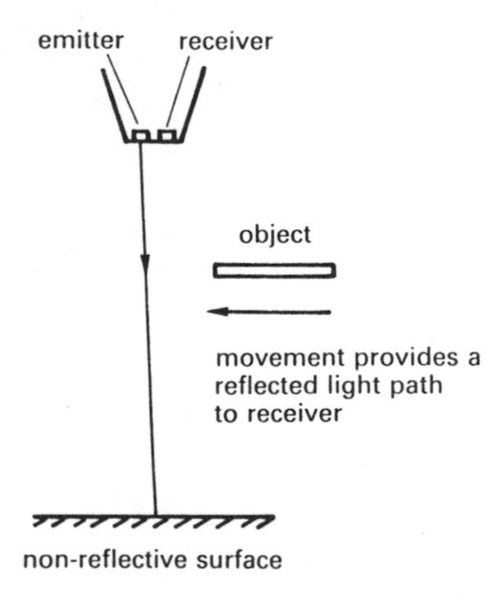


Figure 6.3 Diffuse reflective technique

### *6.1.2 Television cameras*

The common feature of all types of television camera is their production of a *video* output signal via the use of a *raster scan* to obtain the electronic image. The operation of an idealised raster scan comprises the movement of a spot rapidly across the picture and slowly downwards, as shown in figure 6.4a. The corresponding idealised video signal is given in figure 6.4b. A practical video signal is more complex than this. Firstly, '625' line cameras actually produce two '312.5' line *interlaced* scans as seen in figure 6.5. Other complications arise from the finite flyback time, a finite delay between successive interlaced scans, extra pulses to synchronise the X and

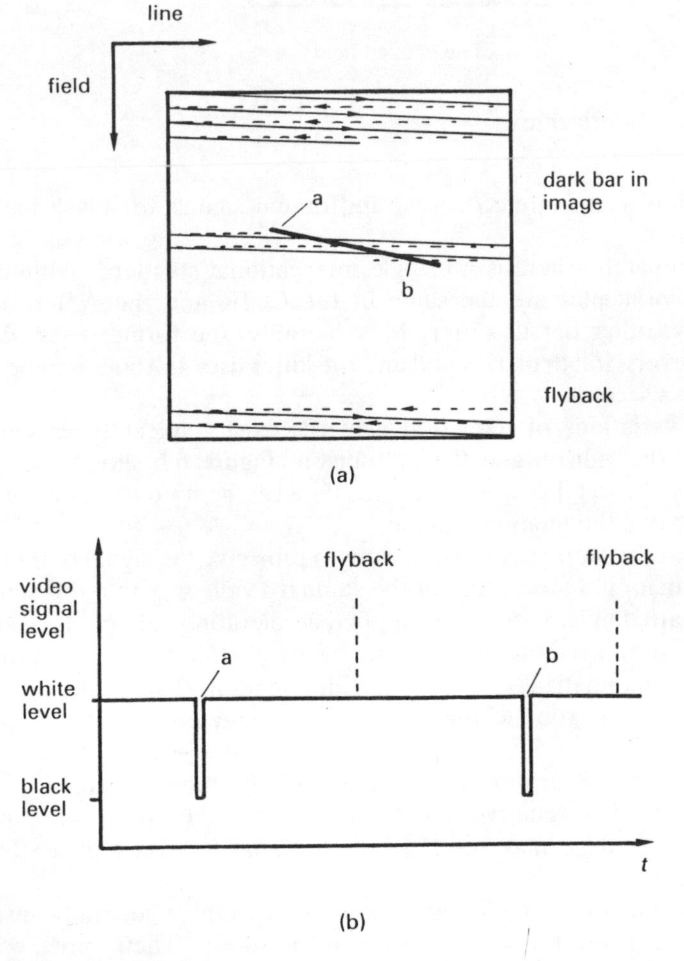


Figure 6.4 (a) Raster scan (simplified), (b) video signal (simplified)

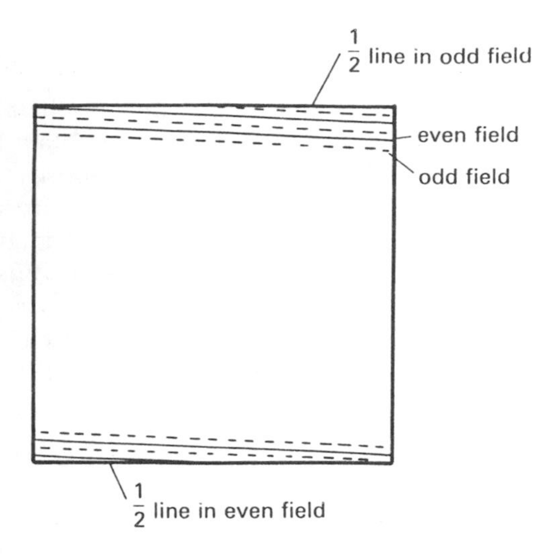


Figure 6.5 Interlacing

Y direction scanning electronics, and datum signals for black and white levels.

Unfortunately, there is no single international standard. Although the scanning principles are the same in the CCIR and the American EIA standards, other details differ. Most notably, the former uses 625 lines scanned every 1/25th of a second and the latter uses 480 lines scanned every 1/30th of a second.

Many variations of television cameras exist, typical types being the orthicon, the vidicon and the plumbicon. Figure 6.6 shows the spectral responses of two types of sensor. This data can be used to advantage when designing the illumination scheme.

The *transfer characteristics* of a camera tube give the signal output versus the illumination. One feature of the standard vidicon is that this sensitivity may be varied by modifying the applied accelerating voltage. It is common for vidicon cameras to be supplied with a circuit which controls the sensitivity automatically to suit the illumination. This feature can be a mixed blessing in robotic applications and it may be better not to use the option.

*Image retention*, or *burn-in*, is the ability of a tube, particularly vidicon types, to hold a semi-permanent image of bright parts of a picture. Long-term damage may result from continual focussing of a stationary bright image.

The size and mass of television cameras make them generally unsuitable for mounting on the end effector of a robot. Their most common application is as stationary cameras located over work areas and as static

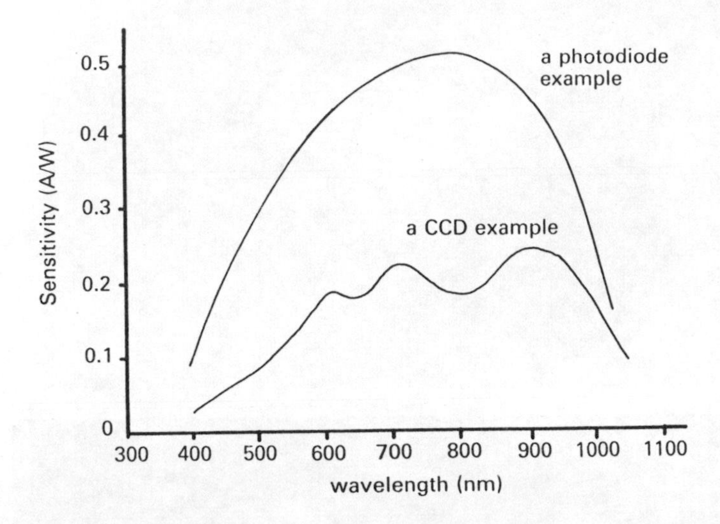


Figure 6.6 Spectral responses — after Batchelor *et al.* (1985)

visual inspection devices. Once a satisfactory image has been obtained, the next problem is to take the continuously varying video signal and put a representation of it into the memory of a computer for subsequent image processing. A common way of achieving this is to use a commercially available *frame store*. This samples the video signal and then performs analogue-to-digital conversion of the samples.

The sampling rate determines the *spatial resolution* of the image. Each sample value is a measure of the illumination at a single point, a *pixel*, in the image. The image is now in digital form so it can be fed into blocks of dynamic RAM on the frame grabbing board. This memory may then be read directly from a program on the host computer as an array of pixel intensity values.

### *6.1.3 Solid state cameras*

Solid state sensors were originally introduced as cumbersome arrays of large photodiodes. The evolution of integrated circuit technology enabled many elements to be included on a single silicon chip. The scanning circuitry was later added on to the chip, obviating the need for large numbers of external connections. Solid state cameras are available in one and two dimensional forms and are of several types: photodiode arrays, charge-coupled devices (CCDs), charge injected devices (CIDs), and dynamic random access memory (DRAM) devices. Most have a spectral response typical of silicon, as shown in figure 6.6.

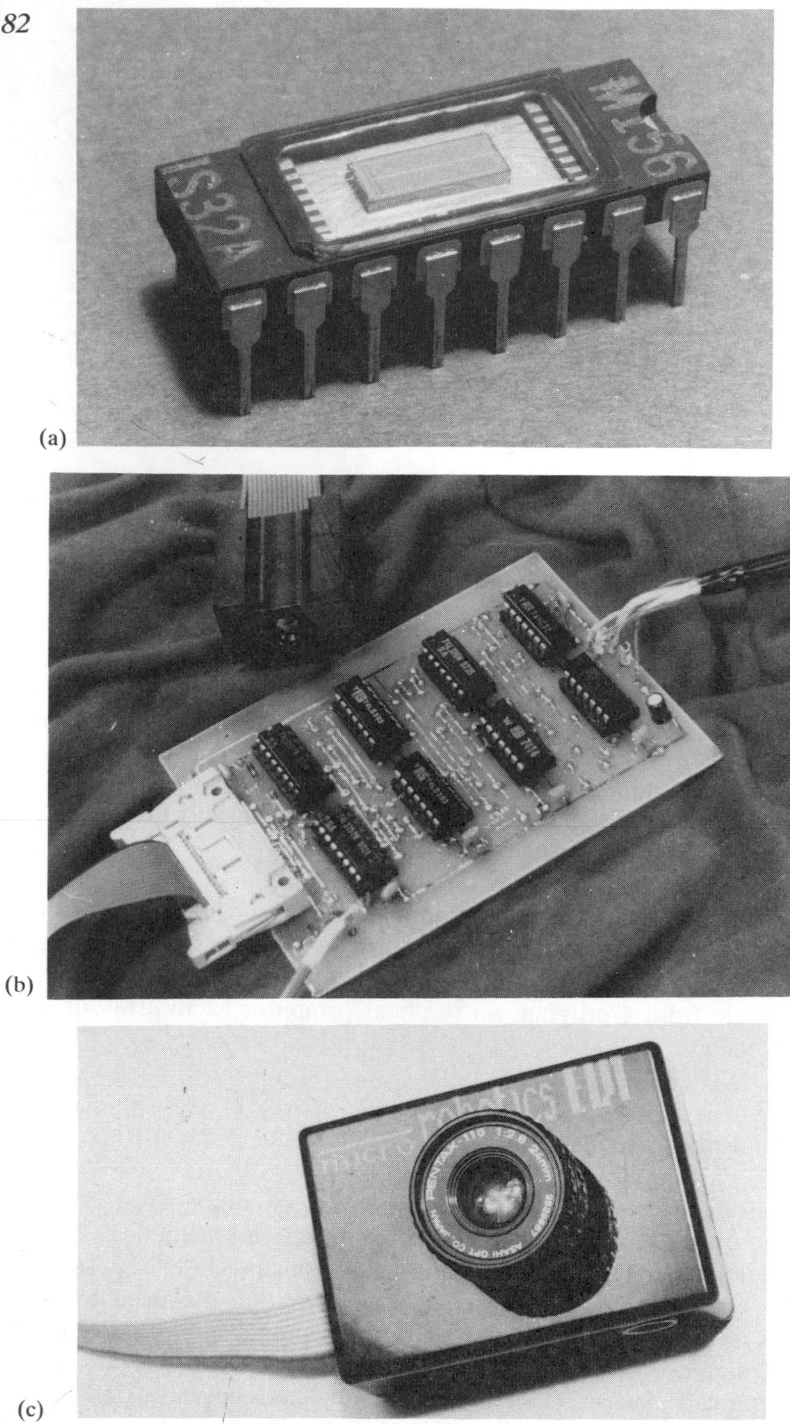


Figure 6.7 (a) DRAM chip. (b) Hull DRAM camera plus drive electronics. (c) Micro-Robotics DRAM camera

In the *photodiode array*, an array of metal oxide on silicon (MOS) transistors can be used to sequentially connect and disconnect each photodiode element with a common video line held at a constant voltage. During the connection time, the capacitance associated with the photodiode is 'reset' by charging it up to close to the video line voltage. Light falling on the photodiode element creates charge carriers which cause the capacitor to discharge. After disconnection, the discharge will increase according to the light intensity and the length of the disconnection time. When the video line is reconnected to the element, current flows from the line to the element. This current can be used as a measure of the voltage drop and hence of the light falling on the element while disconnected from the video line.

In a *charge-coupled device*, arrays of electrodes are placed close to the surface of the silicon chip. As light falls on each element, small amounts of charge are created due to the release of free electrons in the semiconductor. In a simple form of CCD, an analogue shift register is used to transfer the charges from element to element until they appear sequentially at an output, giving a varying analogue voltage which can be sampled and digitised. If the shifting time is not negligible compared with the exposure time, then the charges can be appreciably modified by the incident light during the shifting operation. An alternative approach is to separate the two parts, transferring the charge from each light-sensitive element to a CCD shift register which is shielded from the light. CCD linear arrays of between 256 and 4096 elements are available. Two-dimensional arrays can be obtained in a variety of array sizes. A typical array is the Fairchild CCD211 which has a 244 × 190 element array and operates at 7 MHz.

*Charge injected devices* are similar to the CCD devices. The charges are now kept at the imaging sites and digital scanning circuitry is required. Non-destructive readout is possible and the scanning circuitry can in theory be designed for random access reading.

The *dynamic RAM* type sensor is based on dynamic RAM technology. A typical device, the IS32 optic RAM chip from Micron Technology Inc., is based on Micron Technology's MT4264 64K dynamic memory chip, shown in figure 6.7a. When light strikes a particular memory element, the capacitor associated with that element discharges from the +5V rail to ground. The amount of discharge which occurs is proportional to the intensity of the light and the duration of the exposure. The voltage is measured with respect to a threshold and can be read during the refresh cycle which recharges the capacitor. The elements in the memory array are not linearly arranged and so a small decoding circuit is used to unscramble the rows and columns. An image can then be obtained by merely clocking through each successive row of memory, reading and refreshing each individual element. The elements are arranged in two 32K blocks separated by a 120 micron gap which is an optically blind area. This must be tolerated by the application software or, alternatively, only one of the blocks may be used. Each block corresponds to an array of 128 × 256 pixels.

The chip may be packaged by the user, as can be seen in the Hull camera of figure 6.7b. It may also be purchased as part of a commercial camera system, the EV1 'Snap' system (Micro-Robotics Ltd, 1984), complete with interfaces to a variety of personal computers and a software library to run on them. This camera package is shown in figure 6.7c and houses the decoding circuitry. A detailed discussion of these cameras and how their properties may be used to best advantage may be found in Taylor, G. E. *et al*. (1987).

The major advantages of solid state sensors are that they are rugged, light, compact and cheap. They may be readily used as static overhead cameras or may be gripper-mounted. The cheapness of DRAM types means that it is economic to have several cameras distributed about the gripper and workstation, each concentrating, on particular objects or features of an object (Taylor, P. M. and Bowden, 1986).

Several problems may occur with solid state cameras. Non-uniformities in construction cause variations in pixel sensitivity. Blooming may occur when excessive light is applied to an element on CCD, CID and DRAM types. Excess charge from one element can spill over into the next and thus a very bright spot will spread over the image if the exposure time is too great. Crosstalk may also occur. The areas between the nominal light-sensitive areas may also have some sensitivity to light. Photoelectrons from these areas may be collected by the elements at either side. Nevertheless, with sensible control of lighting and exposure time, these latter problems should become negligible. The advantages, as listed above, make them justifiably popular for robotic visual sensing applications. They are not as highly developed as television tube cameras and improvements in quality and cost are likely.

### 6.1.4 Camera location

The location of cameras for robotic applications may be conveniently classified into one of two sets: static cameras and cameras mounted on moving devices such as robot grippers.

Static cameras, which may be of the television tube or solid state types, may be mounted above a work area in order to locate and/or inspect components as shown in figure 6.8. Careful setup and calibration are necessary. The image co-ordinates of the object as seen by the camera must be correctly transformed into world co-ordinates for use by the robot. Probably the best way of doing this is for the robot to be programmed to

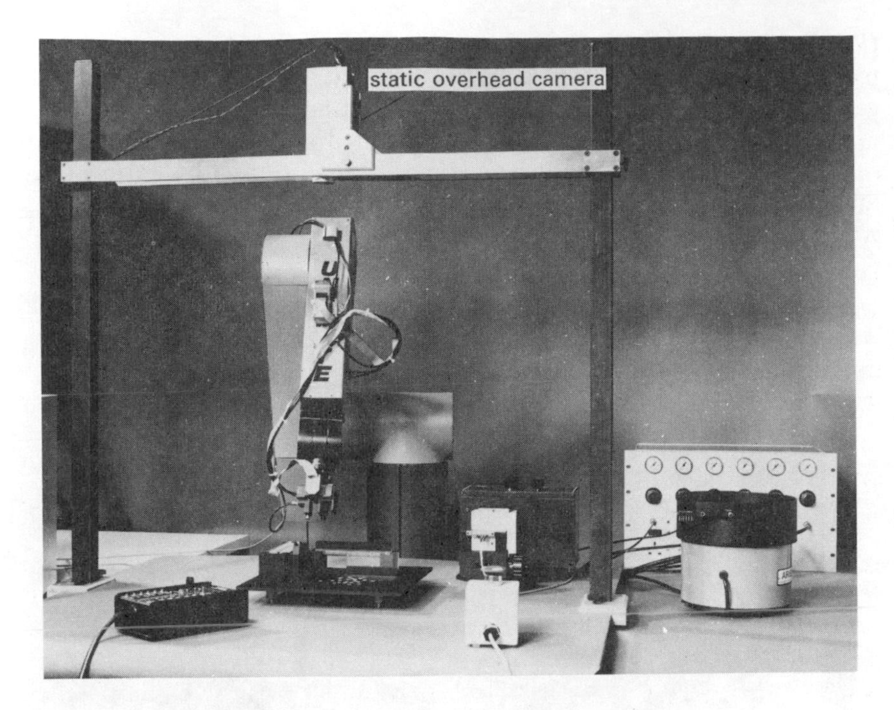


Figure 6.8 Overhead static camera

move a marker under the camera during the workcell initialisation. The transformation may then be found directly from the readings of image and robot co-ordinates. This avoids reliance on the calibrations of the robot and camera given by the manufacturers.

Static camera location is particularly useful for verifying component locations while robots and other machines are performing other operations. Image processing may then be undertaken in parallel with these other parts of the task. However, this approach cannot be used when the object is obscured by parts of the robot or other machinery.

Gripper-mounted cameras, nowadays almost always of the solid state type, are more versatile since they are moved with the robot. They can be used to get close-up images of complete objects, or features of objects, just prior to, or even during, their manipulation by the robot. They are less easily obscured than overhead cameras.

It can be advantageous to combine both approaches. Figure 6.8 shows a static overhead vidicon camera used to find the approximate locations of silicon pellets on a table. Once a pellet has been found, the robot moves over the pellet location prior to picking it up by a vacuum-nozzle. A DRAM type gripper-mounted camera, similar to that shown in figure 6.7b, is then used to locate the pellet accurately and to check for pellet circularity

(Burgess *et al.*, 1982). Figure 6.9 shows three DRAM type cameras mounted on a gripper. Each camera is aimed at a feature of an object and can therefore be used to find the feature with high accuracy.

The main disadvantage of gripper-mounted cameras, particularly for servoing operations, is the increase in task time caused by extra movements and pauses of the robot.

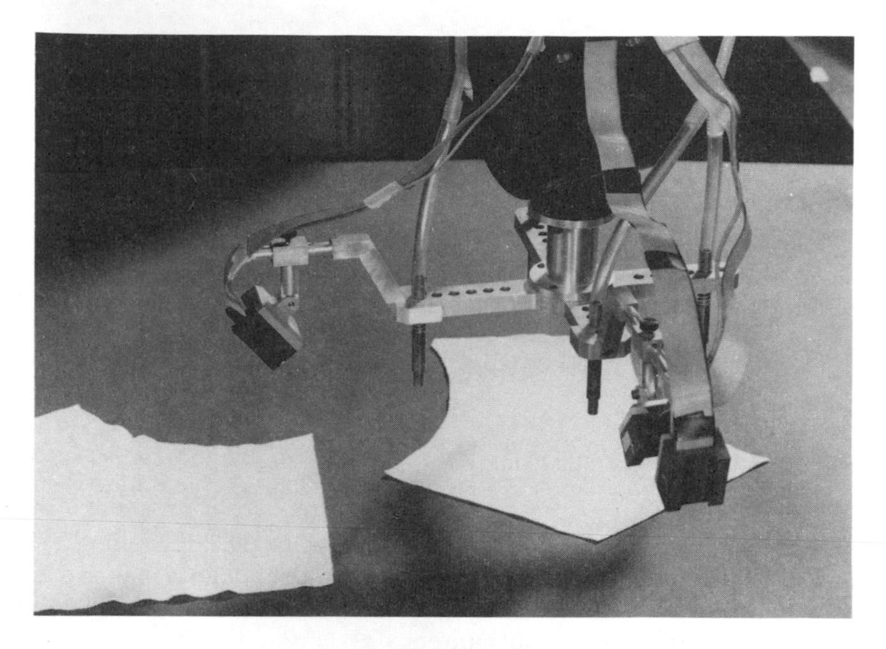

Figure 6.9 Gripper-mounted DRAM camera

### *6.1.5 Lighting*

Good design of the lighting arrangements is often the most important part of a vision system. Good lighting will ensure that an image of sufficient quality is obtained under all operating conditions. Expense incurred in obtaining a high-quality image can save many times this amount in camera and processing costs. Factors to be considered are the type of light source used, how light is applied to the scene, and how the light from the scene is passed to the optical sensor.

Ambient light sources are the simplest to use, requiring little or no additional work. In most factory installations, this means light from fluorescent tubes or in some cases sunlight. In the first case, the camera or image processing software must be able to cope with disturbances such as faulty tubes, shadows and reflections. In the second case, weather changes are significant. As far as possible, it is best to impose known and fixed lighting on the scene to minimise these types of disturbance.

Factors influencing the choice of light source include the colour spectrum of the source; ideally, it should match the camera sensitivity which should peak outside the ambient spectrum where disturbances are likely to occur. Flicker from the use of mains electricity can cause problems, particularly when the camera scanning circuitry is not mains-synchronised. The use of DC or high-frequency sources overcomes this difficulty.

The relative positions of camera, light source and object can be optimised according to the task being undertaken. Batchelor *et al.* (1985) provide an excellent set of examples and case studies showing how a little thought at this stage can be used to good effect.

The most commonly used technique is toplighting, shown in figure 6.10, where the reflected light is used. If the object is not flat then parallax errors may result. Unwanted shadows may be cast from a single source, so multiple sources or light rings may be used. Many image processing algorithms use only the outline of an object. Backlighting, shown in figure 6.11, results in a high contrast image, provided that the surface on which the object rests can be kept sufficiently clean to avoid background 'noise'. Side lighting may be used deliberately to enhance surface detail on the object. The quality of the image may also be improved by adding polarising filters to the light source and/or camera.

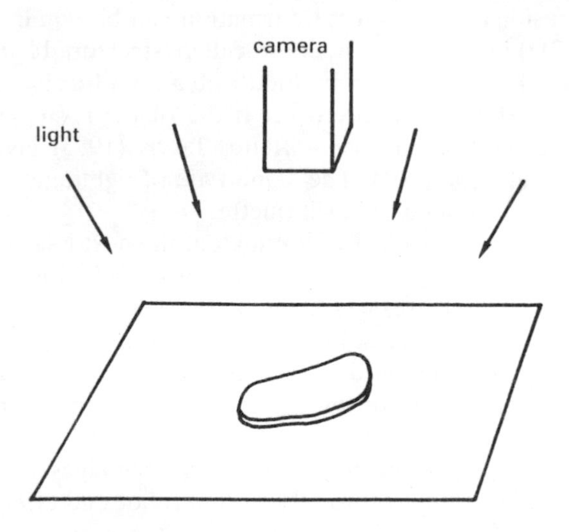


Figure 6.10 Toplighting

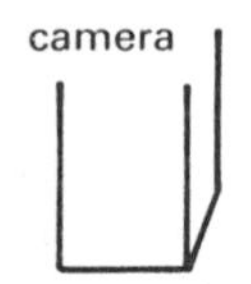


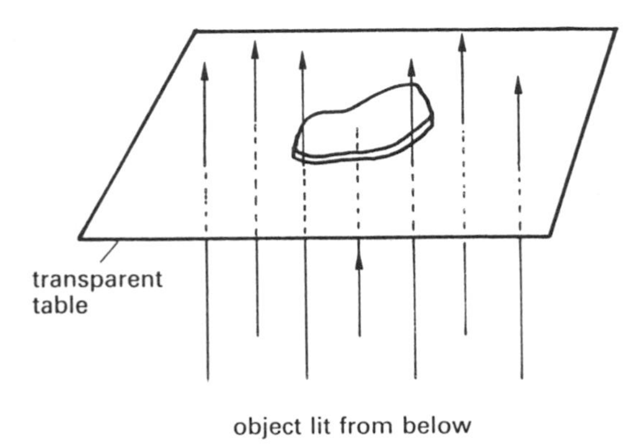


Figure 6.11 Backlighting

Three-dimensional (3D) object information can be obtained from two-dimensional (2D) images by using the so-called 'structured light' approach. A high-intensity light source is combined with a cylindrical lens to project a plane of light as shown in figure 6.12. If the object is stepped along the conveyor, a 3D model may be built up. Jarvis (1983) gives details of various lighting arrangements. The same type of lighting can be twinned with a linear array to build up a silhouette.

Once the object is satisfactorily illuminated, the next issue is how best to pass the image of the scene to the vision sensor. Television cameras and solid state cameras use lenses which can be chosen to give a sharp image of the object inside an appropriate field of view at an acceptable object-to-camera distance. Non-standard lenses, for example cylindrical, may be substituted, deliberately distorting the image to suit the particular image processing application.

Fibre optics provide a means of transferring images from otherwise inaccessible locations, but not usually from a robot end effector since the continual flexing of the fibres can lead to breakages.

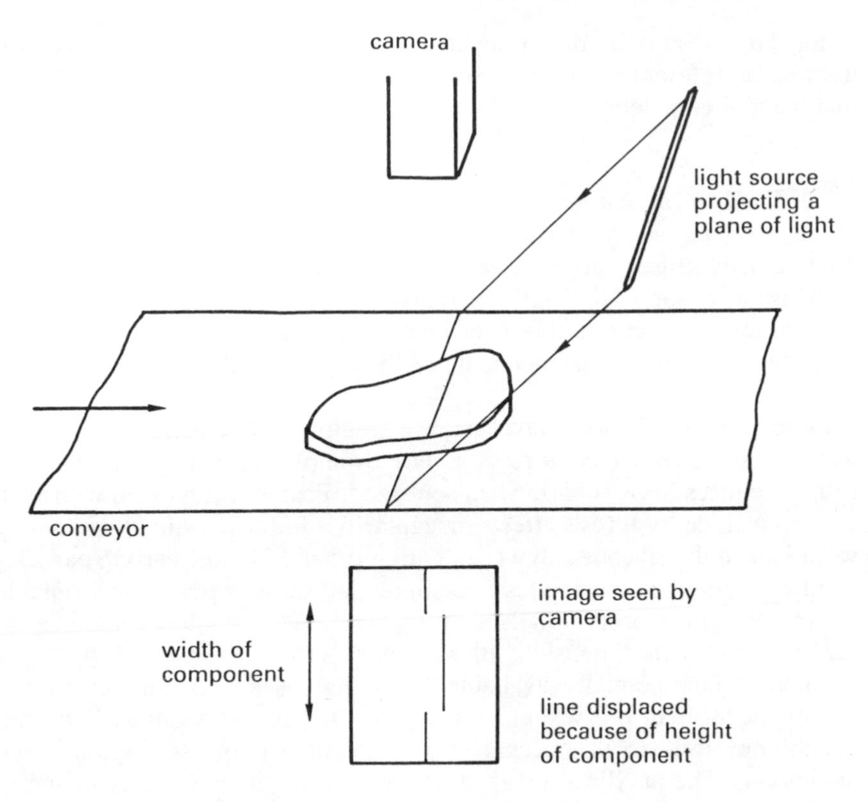


Figure 6.12 Structured light

## 6.2 Touch and tactile sensors

Touch and tactile sensors require physical contact to be made between the object and the sensor. The convention used here (Harmon, 1982) is that 'touch' refers to simple contact for position or force sensing at one or a few points only. 'Tactile' refers to graded sensing of positions or forces in an array.

### *6.2.1 Touch sensors*

The most widespread touch sensor is the simple switch. Switches are cheap and reliable, better than $10^7$ operations before failure being quoted for a typical microswitch. They are frequently used to detect when robotic arms have reached close to their allowable limits of operation. They are also used as reference positions. For example, during the initialisation of a robot, an axis may be driven against a microswitch; this position is then used as a reference base for subsequent position measurement using incremental encoders.

### *6.2.2 Tactile array sensors*

Tactile array sensors are in a very early stage of development compared with vision sensors. They have been proposed for use in contour examination, surface inspection, object recognition, grasp error compensation and assembly monitoring. Each element of the tactile array is called a *tactel* or *forcel*.

Piezo-resistive devices have array elements with electrical resistance which changes under compression. For example, carbon-loaded silicone rubber reduces in resistance when squeezed. Problems reported with such sensors include hysteresis effects, irregularities in the sensitivity caused by variations in the silicone rubber, and fatigue caused by the carbon particles creating points of local stress. Commercial sensors are now available (Barry Wright Corp., 1984) with typical linear resolutions down to 1.25 mm and with dimensions 10 × 20 mm.

Another commercially available technique uses a compliant rubber membrane which presses against an acrylic sheet. Contact between membrane and acrylic occurs when an object presses against the membrane. The acrylic sheet is illuminated along its length, as shown in figure 6.13. The total internal reflection of light is broken at the contact point, and light is then scattered out of the sheet and passed to a CCD sensor. Typical resolutions of 0.1 mm are achieved over a 16 × 25 mm surface. The construction is compact enough to be incorporated into tactile fingers for a gripper as can be seen in figure 6.14 (Mott *et al.*, 1984).

Other techniques using piezo-electric, magneto-resistive, capacitative, and inductive principles are under investigation in research laboratories.

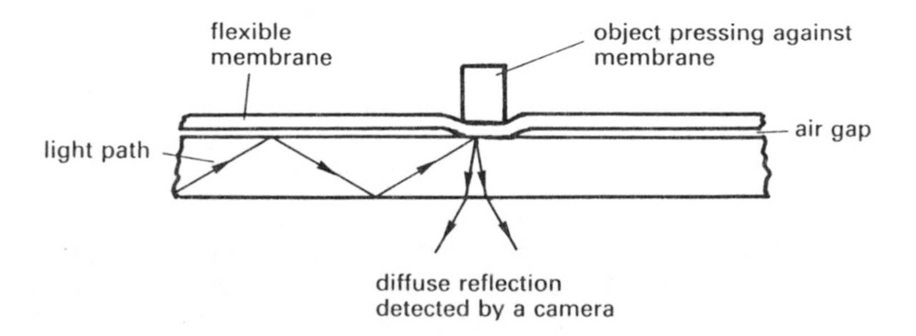


Figure 6.13 Principle of Aber tactile sensor

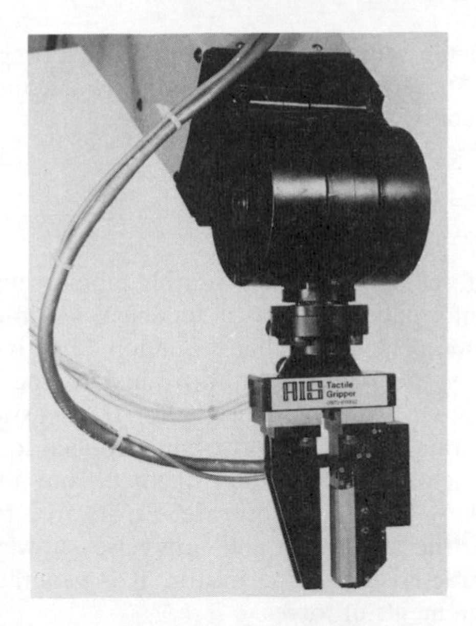


(a)

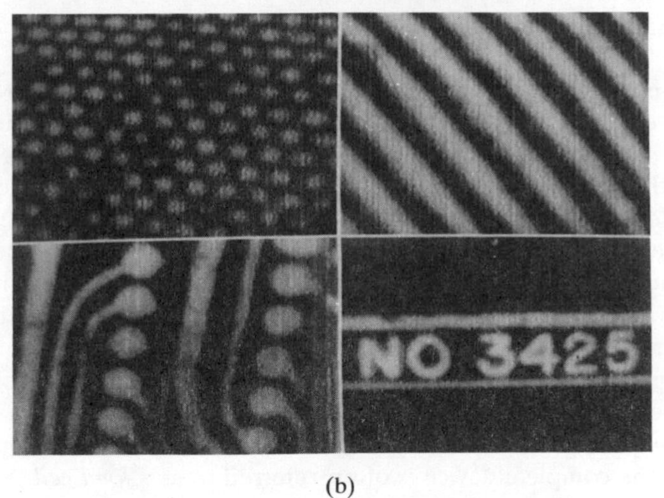


(b)

Figure 6.14 Aber tactile fingers:
(a) tactile gripper; (b) tactile images

| handfile surface | ribbon cable |
|---|---|
| pcb tracks | raised lettering on plastic pipe |

[courtesy of Department of Computer Science, University College of Wales, Aberystwyth]

One of the first tactile sensors was developed at Nottingham University (Pugh *et al.*, 1977). It used a 16 × 16 matrix of rods at a 5 mm pitch passing through inductive coil windings.

## 6.3 Force sensors

Measurements of force are used for assembly monitoring, as a means of obtaining tactile information, and as a feedback for improved dynamic control. If a DC motor is used, then equation (3.6) indicates that the torque produced by the motor is proportional to the current passing through it. This torque will eventually, after transmission through any gearbox and drive train, result in a force being applied to an object by the end effector. In practice, the derivation of force from a measurement of motor current is likely to be too inaccurate. Errors arise from unmodelled frictional effects in the drive train and imprecise knowledge of the true torque/current characteristics of the motor. It is generally better to use more direct measurements of forces.

### *6.3.1 Strain gauges*

As wire is stretched, it becomes thinner and longer and thereby has increased electrical resistance. If such a wire is bonded on to the surface of a bar and aligned with the axis of the bar, then its resistance can be used to measure the extension of the bar when a load is applied along its length. If the force/extension characteristic of the bar is known, then this resistance can be used as a measurement of the applied force. In practice, the wire is wound back and forth to exaggerate this effect, and is combined with other gauges in a bridge network in order to nullify temperature effects. A typical arrangement is shown in figure 6.15.

Strain gauges are also supplied ready-mounted on carefully manufactured blocks or beams which can be inserted into the wrist of a robot. Again, multiple gauges can be used to reduce sensitivity to temperature changes. The complete device is often referred to as a *load cell*. Since the bridge network produces low voltages (ideally zero when balanced with all cells equally loaded), low drift, temperature-insensitive amplifiers are employed to give a more useful output voltage.

## 6.4 Other sensors

A common robotic requirement, especially with mobile robots, is to know the distance to an object. Tactile sensing requires the sensor to be brought

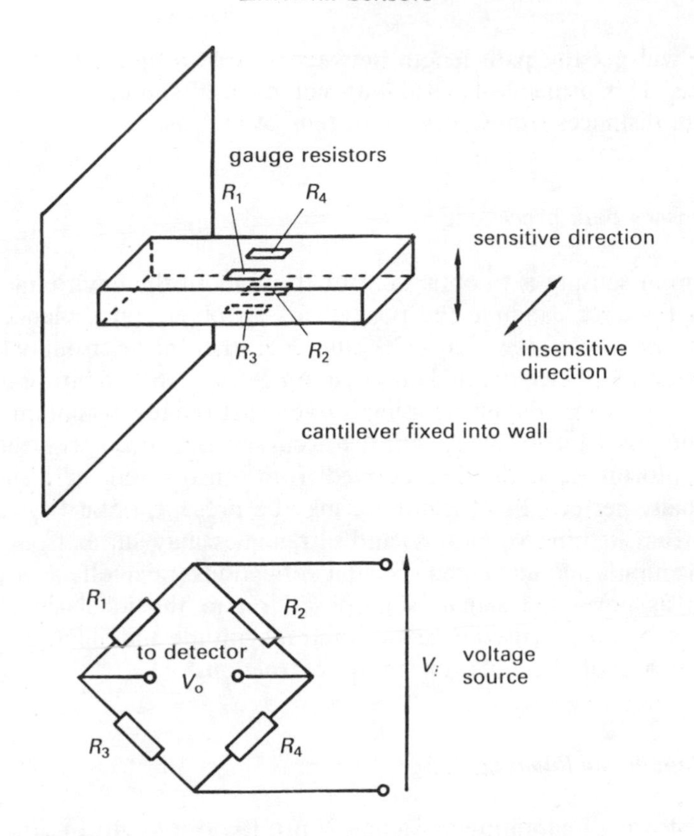


Figure 6.15 Strain gauge bridge

into contact with the object, and this need for movement is often very wasteful of time. Possible vision techniques which may be used are: structured light, stereoscopic images using multiple cameras, measurement of the time-of-flight of a reflected light beam, measurement of the phase shift in a reflected laser beam, and use of *a-priori* knowledge of the outline size of the object at different ranges. However, vision sensing is often impracticable. Other approaches to *range finding* are often simpler and are in more widespread use.

### *6.4.1 Ultrasonic sensors*

A pulsed sound wave propagated through air is reflected when it hits the interface between the air and an object. Suppose we measure the time taken between the emission of a sound pulse from a source, to its receipt by a sensor close to the source. If we decide this time by the speed of sound in air we will get the path length between source, object and back to the receiver. This principle is used with commercial source/receiver pairs to measure distances from a few cm to tens of metres.

## 6.5 Sensory data processing

The aim of sensing is to obtain information about the environment. Such information, for example the position of an object on a plane, may be encapsulated in a few, say 32, bits. Yet the image from which this information is to be extracted may contain $256 \times 256 \times 8$ bits of data. The purpose of sensory data processing is to extract the few bits of information from the many bits of data, with the accuracy and speed required for the task. Unfortunately, the data derived from sensors and their digitisers is not usually perfect. Electrical noise may be present, optical systems may suffer from lighting variations, and ultrasonics may suffer from external acoustic inputs and unwanted specular reflections. Such effects are usually treated as unwanted signals superposed on to the ideal signal. Signal conditioning is often used to reduce their magnitude and to modify the data into a form more suitable for feature extraction.

### *6.5.1 Signal conditioning*

The best way of conditioning signals is not to do it at all! In other words, the sensing system should be designed so that the sensor and its associated electronics produce near-ideal signals. This requires thought being given to the environment in which the sensing system is working. Sources of interference should be identified and, if possible, eliminated at source. Alternatively, shielding should be provided here or at the sensor and its associated electronic circuitry.

The extraneous noise is often in a frequency range outside that of the wanted signal. In these cases, the use of low or high pass filters may eliminate the noise and yet leave the required signal substantially unchanged. An example here might be to extract a slowly varying force signal from a load cell subject to high-frequency vibrations. Another would be to filter an image to remove isolated pixels.

### *6.5.2 Feature extraction*

Perhaps the most commonly required piece of information is the position of an object within an image. A simple way of determining this is to find its centroid. The task is greatly simplified if it can be guaranteed that there is only one object within the image. The centroid measure is invariant with object rotation provided that the pixels are equi-spaced in each of the $x$ and $y$ directions.

Orientation is another common parameter to be measured. The example given in section 7.5 describes how a rotation table combined with a template matching operation can be used to achieve a desired orientation. Other schemes are based on computing the orientation from a single image. One example (Loughlin *et al.*, 1980) characterises an object by the numbers of pixels lying in the radial segments of a circle centred on the centroid of the object (see figure 6.16). These numbers are put in a list and compared with the list obtained from a taught image in a given reference orientation. The numbers in the list are shifted along (corresponding to a rotation of the object) until the best match is obtained.

Other measures are based on finding the orientation of simpler, related shapes, for example, the orientation of the major axis of the smallest enclosing rectangle, or the orientation of the major axis of an approximation ellipse (Agin, 1985).

Quite often, it is only necessary to find the position of features such as corners and edges. Various edge detection algorithms can be used, a typical one being the Sobel gradient operator shown in figure 6.17.

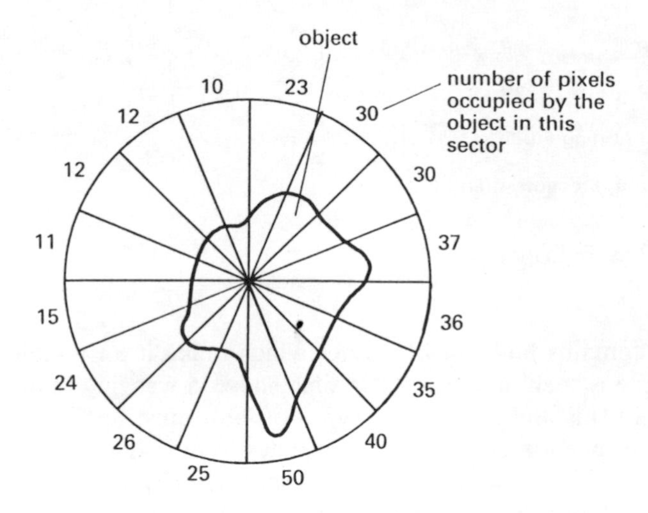


Figure 6.16 Orientation using circular segmentation

### *6.5.3 Inspection*

Inspection is generally a much more difficult task than that of extracting the features needed for handling an object. It is now necessary to find out if

| | | |
|---|---|---|
| $P_{y-1,x-1}$ | $P_{y-1,x}$ | $P_{y-1,x+1}$ |
| $P_{y,x-1}$ | $P_{y,x}$ | $P_{y,x+1}$ |
| $P_{y+1,x-1}$ | $P_{y+1,x}$ | $P_{y+1,x+1}$ |

| | | |
|---|---|---|
| −1 | 0 | 1 |
| −2 | 0 | 2 |
| −1 | 0 | 1 |

Sobel horizontal mask (h) used to find horizontal edges

| | | |
|---|---|---|
| −1 | −2 | −1 |
| 0 | 0 | 0 |
| 1 | 2 | 1 |

Sobel vertical mask (v) used to find vertical edges

$$h = (P_{y-1,x+1} + 2P_{y,x+1} + P_{y+1,x+1}) - (P_{y-1,x-1} + 2P_{y,x-1} + P_{y+1,x-1})$$

$$v = (P_{y+1,x-1} + 2P_{y+1,x} + P_{y+1,x+1}) - (P_{y-1,x-1} + 2P_{y-1,x} + P_{y-1,x+1})$$

$$\text{Gradient magnitude} = \sqrt{h^2 + v^2} \approx |h| + |v|$$

$$\text{Gradient direction} = \tan^{-1}\left(\frac{v}{h}\right)$$

Figure 6.17 Sobel operator

the object contains faults or blemishes which make it unsuitable for use. One example is seen in figure 6.18 which shows a weaving fault in a shirt collar panel. This and many other types of fault must be detected in any automated inspection call. The difficulty lies in devising a strategy which reliably produces acceptable results in a satisfactorily short time and at a low cost. Best results are obtained by careful combinations of lighting, sensing techniques and algorithms (Batchelor *et al.*, 1985).

Most implementations of inspection systems use standard or customised serial computers with, perhaps, some image analysis functions evaluated using analogue hardware or special-purpose integrated circuits. Braggins and Hollingum (1986) give a survey of commercially available systems for both inspection and handling tasks.

Figure 6.18 Shirt collar weaving defect

## 6.6 Sensory task control

Once information from the sensors and sensory data processors becomes available, it is necessary to augment the features of the language used to describe the task. In this section we will assumed for simplicity that the task may be described and executed locally within the robot's own controller. In more extensive tasks it may be more appropriate to use a different structure, as discussed in chapter 7.

### *6.6.1 Language requirements*

The program which describes the task must be able to interrogate the sensor or its data processor. Simple sensors such as switches may be interfaced to the robot via a set of binary input lines. The language must then have statements which can interrogate these lines, for example in the VAL language (Unimation Inc., 1985)

```
WAIT SIG(1001)
```

halts program execution until signal 1001 is turned on.

Sometimes these lines may be bidirectional, that is input–output (I/O) lines. A statement is then needed to define whether the line is to be used to input or to output data; for example, in the AML language (Taylor, R. H. *et al.*, 1983)

```
DEFIO(21, 0, 1, 0, 1)
```

is interpreted as follows. The first parameter, 21, indicates the I/O channel number and the following 0 defines it to be for input. The next 1 indicates that an open contact will yield a −1 and an open contact a 0. The following 0 and 1 mean that the input information will be of one bit in width and will start at bit 0 of the input/output word.

Similar statements may be used to output a bit of data to one line. For example, to switch on a vacuum pump connected to line 3, the following type of statement might be used in VAL:

```
SIGNAL 3
```

More sophisticated sensing requires the transmission of complete bytes of information, perhaps sending the forces sensed by a group of six strain gauges. Using the AML language, this information may be transmitted as a set of six real numbers by means of a command like

```
SENSIO (<G1, G2, G3, G4, G5, G6>, 6 of REAL)
```

In many cases this statement would be preceded by a request to the sensor data processor for certain information to be computed from sensor measurements.

Once the sensory information has been received, it must be acted upon. Typically, the program might branch, conditional on the state of one of the input lines, for example

```
IF SIG GOTO 10
```

or execute an error recovery routine:

```
IF SIG THEN GOSUB TURNOVER
```

The sensory information may be put into program variables to use during later movements, for example

```
XCAM = ADC(5)       Get X location from channel 5
YCAM = ADC(6)       Get Y location from channel 6
X = X + XCAM*6      Modify X, with scaling
Y = Y + YCAM*4      Modify Y, with scaling
MOVE X, Y           Move robot to new position
```

An interrupt facility, such as that provided by the REACT command in VAL II, allows the execution of the main program to be halted while an interrupt routine, in this case INTSUB, is executed. This happens when the external binary signal VAR changes from off to on.

```
REACT VAR, INTSUB
```

If the sensor information or the transformation from sensor to robot co-ordinates is inaccurate, then a *servo-loop* may be established. An example would be to move a robot having a gripper-mounted camera over an object, and then adjust the camera's position until the centroid of the object appears at the centre of the image. If, as a result of this, the centroid is not centrally located then the process is repeated, perhaps several times, until convergence occurs. In Pascal, as may be used to control the UMI-RTX robot, this could be represented as:

```
;
;      servo robot until camera central over object
;
       REPEAT;
       Get_centroid (xcam, ycam);
       Robot_xyz[x]:= Robot_xyz[x]+ xcam*6 ; Calc new
       Robot_xyz[y]:= Robot_xyz[y] + ycam*4 ; robot x,y
       Move_arm_xyz (Robot_xyz)           ; and move to it
       UNTIL abs(xcam)+abs(ycam) <1;
;
;      now over object
;
       Save_pos:=Robot_xyz;
```

where xcam and ycam have their origin at the centre of the image.

Visual sensory processing is generally very time consuming. If fixed overhead cameras are used to locate the position of a component for subsequent pick up, it is best if this can be done while the robot is performing earlier actions. This requires *concurrency*; the sensory processing must be performed in parallel with these earlier robot movements. Similar facilities are required when two or more robots are co-operating on a task under one overall controller.

# 7 Workcell Control

Up to now we have considered just a single robot working independently of other equipment, apart from some sensors to provide feedback about the environment. In reality, life is not so simple. The robot is there to perform some operations on one or more objects, so, given a non-mobile robot, these objects must be fed into the working volume of the robot and taken out once the operations have been completed. It is usually necessary to constrain the objects while the robot operates upon them; mechanical restraining devices for holding the objects are classed as *fixtures*. It may be that the robot is used essentially as a programmable loading and unloading mechanism, working in conjunction with another machine, for example a press. The collection of robot(s), fixtures, feeding devices and other machines which are combined to perform one or more operations on the object(s) we will call a *workcell* or just *cell* for brevity.

It is necessary that the operation of the robot be co-ordinated with the other machinery, both to avoid collisions and to ensure co-ordination of operations. This requires some means of communication between the devices in the cell. The particular task being undertaken in the cell may be complex enough to require two or more co-operating robots working in the same space. This further complicates the problems of communications and collision avoidance.

The particular cell may operate as a small island of automation or it may be one of many such cells all working together, say as an assembly line. Although each cell may be working autonomously for most of the time, it will in fact be just performing part of a larger task and again there must be communication links and co-ordination between all cells.

This chapter explores aspects of these topics and culminates in an illustrative example.

F. M. Colls

## 7.1 Workcell mechanisms

These devices are all used to move objects in the working volume of the robot. They may be either simple or more sophisticated programmable devices. In the latter case, note that the operation of several such devices may be co-ordinated to form a distributed robot. A few commonly used devices are considered below.

### *7.1.1 Bowl feeders*

It would make life a lot simpler for robotic workcells if all components could be presented reliably at precisely known positions. For example, in high-volume electronic assembly work, where specialised automated assembly machines are used, it is common to mount resistors on bandoliers. The strips of the bandoliers may be fed into the machinery and one resistor at a time removed from a well-defined position. Other objects may be presented in pallets. However, it may be difficult or just too expensive to structure the component presentation in such a way. Components are frequently supplied in boxes, bags or bins, inside which there may be hundreds of such components jumbled up together. Some means must be used to separate out one component at a time. The most common way of doing this automatically is to use a bowl feeder.

A vibratory bowl feeder is shown in figure 7.1; in this case it is used to feed rivets into an assembly cell. The components are tipped into the body of the bowl feeder which is vibrated up and down and in torsion about the vertical axis. As a result of this, the components travel towards the perimeter of the bowl and up the inclined track which spirals up the inside wall of the bowl. This mechanism ensures a series of components coming along the track. It is advantageous to present such components in a

Figure 7.1 A vibratory bowl feeder

well-defined orientation and position to the robot. This is performed by the *tooling* associated with the bowl feeder. Figure 7.2 shows the guides which form the tooling to ensure that the rivet is presented at the end of the feeder track with its head uppermost. Once one rivet has been removed by the robot, the action of the bowl feeder automatically ensures that another rivet is pushed along the track ready for the next access.

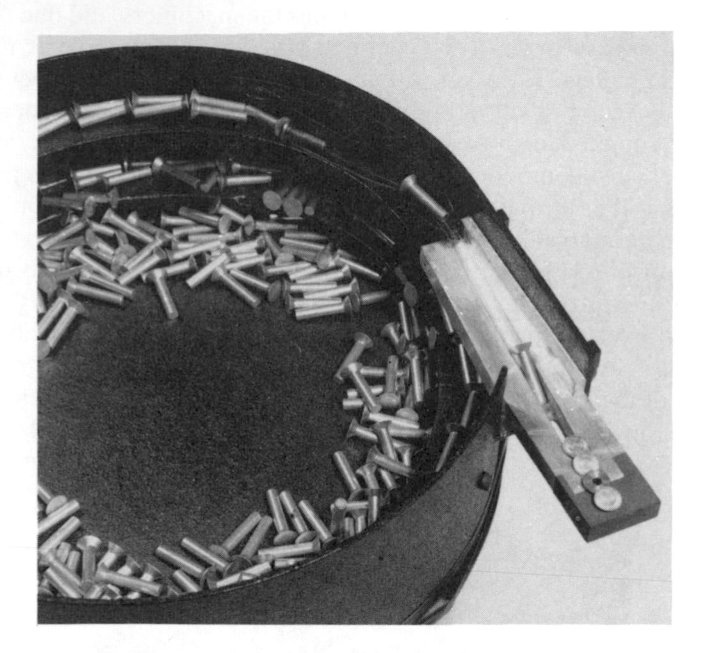

Figure 7.2 Tooling on a bowl feeder

One of the disadvantages of such a bowl feeder lies in the tooling required to give the desired presentation of the object. By its very nature, this tooling tends to be object–specific and must therefore be changed when the component changes. The tooling therefore converts the feeder into a dedicated device, both increasing the cost and reducing the flexibility of the cell. One way of overcoming this problem was suggested in early work at Nottingham University (Cronshaw *et al.*, 1980) and is characterised by using sensors to detect the position and orientation of a component at the end of the track. A very simple guide is used to ensure the presentation of only one component at a time. This *optically tooled bowlfeeder* is shown in figure 7.3. The component, a bicycle brake lug, is pushed across two orthogonal slits each containing a row of optical fibres. The lighting and the feeding mechanism are arranged such that two silhouettes of the object are formed as it is pushed across the slits. Some

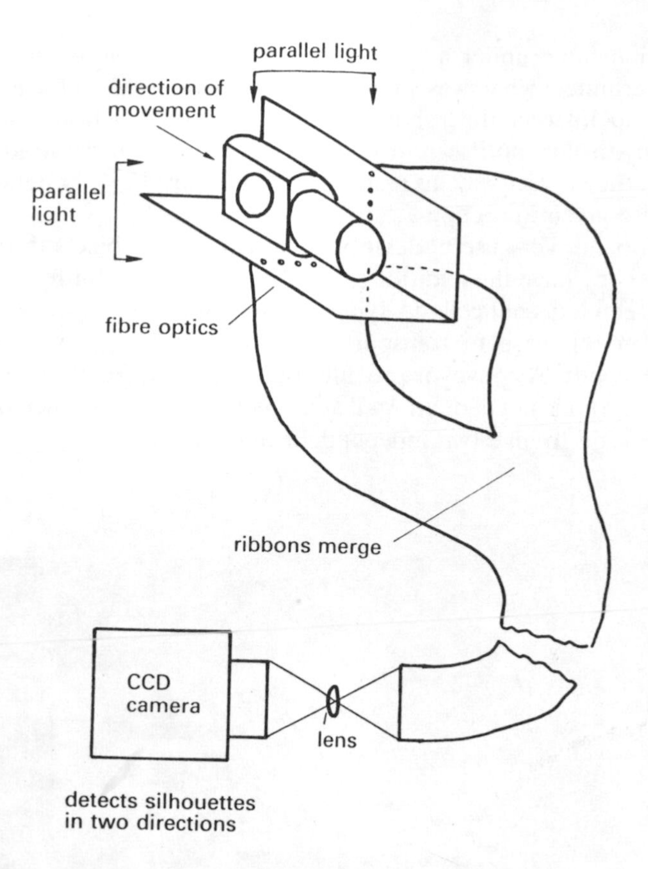


Figure 7.3 Optical tooling — after Cronshow *et al.* (1980)

simple image processing is used to determine in which stable position the component lies.

There are several other types of feeder such as hopper feeders and belt feeders. For details of these and a much more complete coverage of means of feeding parts, the reader is referred to the book *Robots in Assembly* (Redford and Lo, 1986).

It is possible to use a robot to perform the same operation as the bowl feeder. The use of vision in conjunction with a robot for this *bin picking* problem was pioneered by Kelley (Kelley *et al.*, 1983). A camera was pointed into a bin containing crankshafts and the image was processed to find the 'best' possible target crankshaft to try to pick up. Another approach was used at Hull University in the motif application project (Taylor, G. E. *et al.*, 1982). A vacuum sucker was dropped blindly into a bin of embroidered motifs. Whatever was picked up was then dropped on to a rotation table under a camera connected to a simple vision system which determined what was present on the table. This information was then used to instruct the robot to take appropriate action, such as the removal of surplus motifs, in order to finish up with the desired result of one motif, the correct way up ready for orientation. Further details of this example are given in section 7.5.

Conveyors are very useful devices for transporting objects through long distances. They have the additional benefit of providing buffers of components between adjacent cells. A typical conveyor system is shown in figure 7.4, used in this case to transport fabric panels for picking up by the UMI-RTX robot. A conveyor provides one degree of freedom movement. If more freedom is needed, an *X–Y table* such as the one depicted in figure 7.5 can be used to give two independent movements.

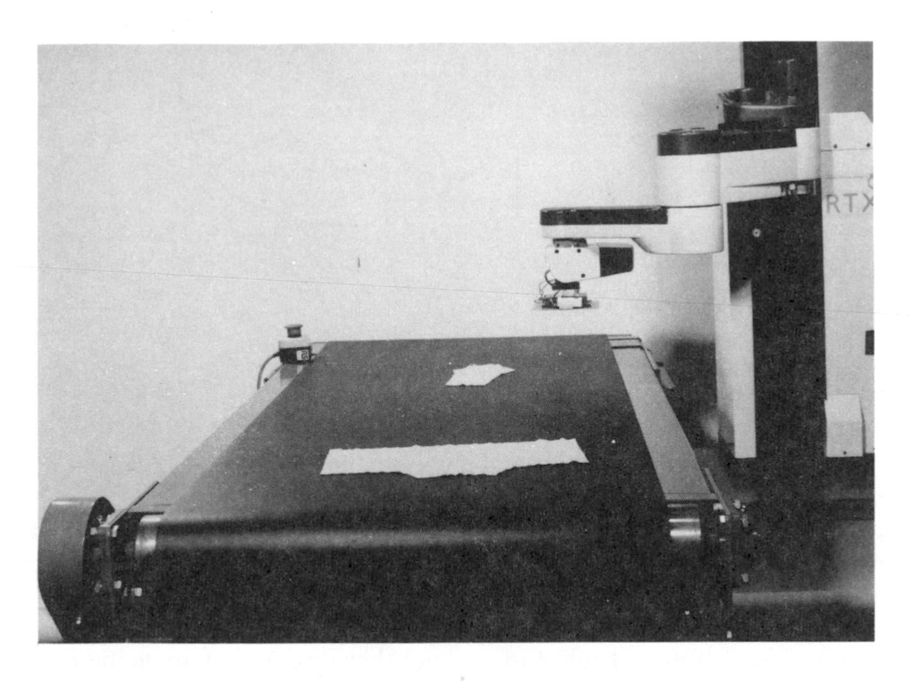


Figure 7.4 A conveyor system

Rotation tables, such as that illustrated in figure 7.6, are used to rotate components into the desired orientation prior to pick up by the robot or feeding into another machine.

Figure 7.5 An X–Y table

Figure 7.6 A rotation table

## 7.2 Fixtures

If two parts are to be joined together, and one is held by a robot gripper, then some means must be provided to hold the other part in place. This is the purpose of *fixtures*. In many cases it is best to provide the fixture with some compliance, especially when there is inadequate compliance in the robot and its gripper.

Compliance along the vertical axis is very helpful when one part is being placed on top of another. Figure 7.7 shows a vertically compliant fixture

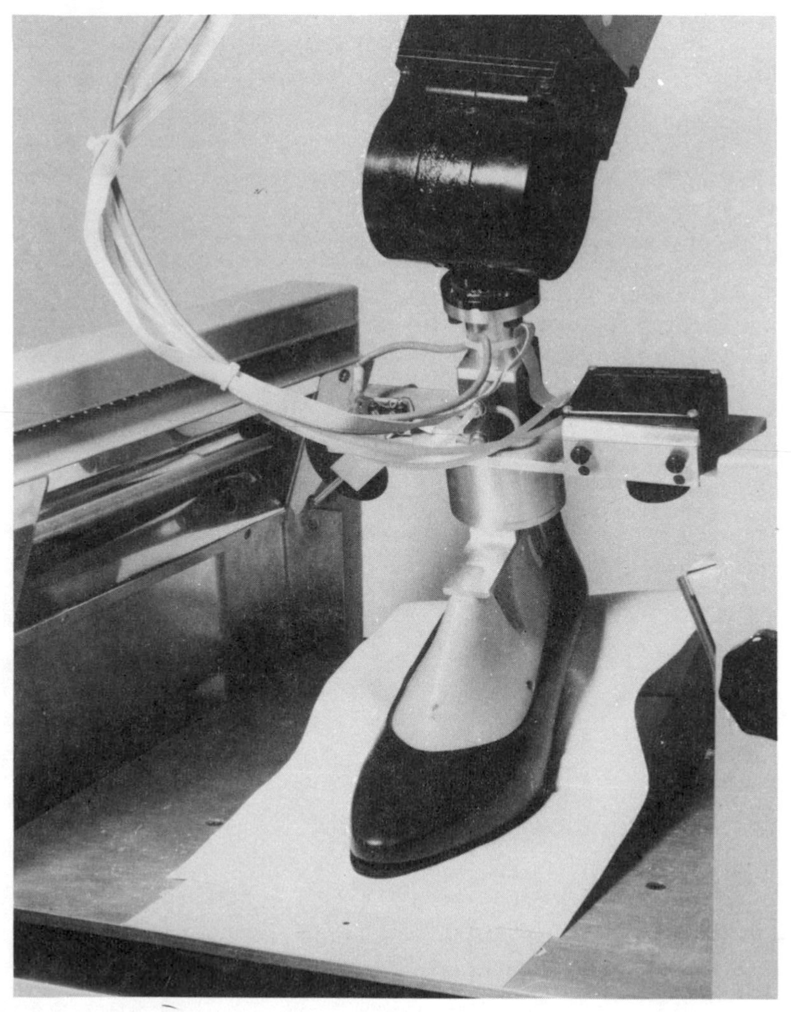

Figure 7.7 Vertical compliance in shoe sole assembly [courtesy of SATRA Footwear Technology Centre]

used to hold a shoe sole while a lasted shoe upper is 'spotted' (a touch bond is made between the two surfaces to which adhesive has been applied). In this case the spotting force can be varied by changing the vertical position of the upper as it presses against the sole on the compliant foam base. This particular fixture also employs an inflatable bag beneath the foam base. When activated, the bag forces the sole upwards to form a good fit against the 3D lower surface of the upper.

Compliance in the horizontal plane is particularly useful when one part has to be inserted into another. In practice, the two parts will not be perfectly aligned. If neither can move during the insertion phase, then failure may occur. The use of a remote centre compliance unit in the gripper, or a robot which is naturally compliant in the horizontal plane (such as some SCARA type robots), are alternative solutions to putting similar compliance in the fixture.

As with feeders, one of the problems associated with fixtures is that they must often be designed specifically for one particular part or family of parts. Again, this adds to the cost of the cell and reduces its flexibility.

## 7.3 Communications

A workcell comprising one robot, a few feeders, fixtures and a vision system may all be controllable from the robot's own controller provided that the manufacturer has provided sufficient interfaces and a rich enough programming language. In this case all devices are physically connected to the robot's controller, probably via standard serial or parallel interfaces. It is likely that such interfaces are optically isolated to minimise damage to the controller caused by undesirable high voltage signals on the input lines. It can be both difficult and time consuming to add extra input–output facilities not built-in to the system. Unfortunately, this difficulty tends to increase as the sophistication of the basic robot system increases. Similar problems arise with computer operating systems. To enter a value into a specific address in the memory of a well-protected multi-user computer is usually non-trivial, whereas for a typical home microcomputer it is very simple.

It may well be that the complete cell is of much greater complexity than that which can be handled by the robot's in-built controller, either in a hardware sense (such as too many input lines of a particular type) or in a software sense (such as the language not containing certain essential features). In such cases, a higher-level controller is necessary and some type of communications structure must be adopted to pass information between the parts of the cell. There is no universally accepted standard to use, and at the moment many large research groups and organisations use their own in-house standards (which may well be based on some proprietary bus system). The structure may well take the form shown in figure 7.8.

The complete task description now resides in the overall cell controller. This may be in terms of a fully fledged robot language of the type described in the earlier chapters, or a standard programming language. It is perhaps more appropriate in such a large system to think of the robot(s), other machines and sensory processors as peripheral devices. Some of the devices are capable of carrying out perhaps one simple action on request, others may contain their own software such that their operation can be changed on command from the overall controller.

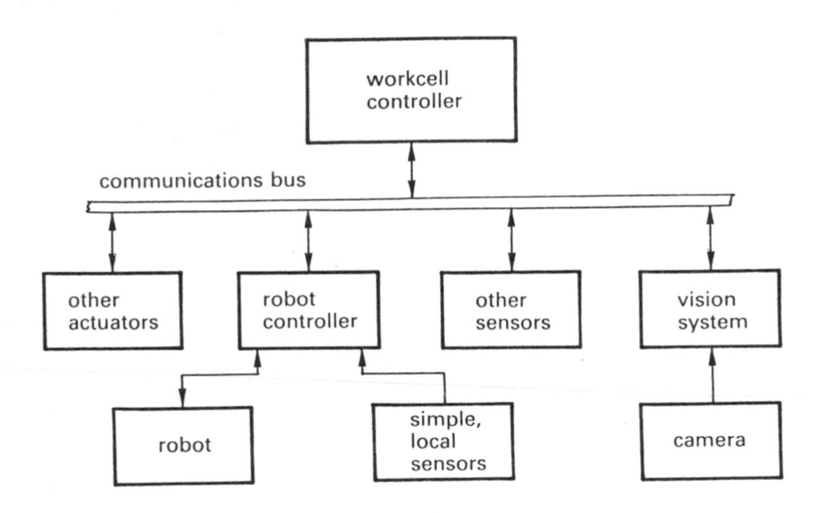


Figure 7.8 Workcell control structure

At a higher level still, this workcell may be working in concert with a series of other workcells as part of a large automated system, say for the assembly of a complete product. One approach is to use another level of controller to co-ordinate all the cell controllers beneath it, as shown in figure 7.9. Communication at this level may be via one of the emerging standards for manufacturing plant such as the Manufacturing Automation Protocol (MAP) (General Motors, 1985). However, MAP coupling is currently expensive and so for this and other reasons, such as the need for fast real-time communications within the robotic cell, the optimal solution is probably some combination of MAP at the high level, and another local area network (LAN) at the low level (Walze, 1986).

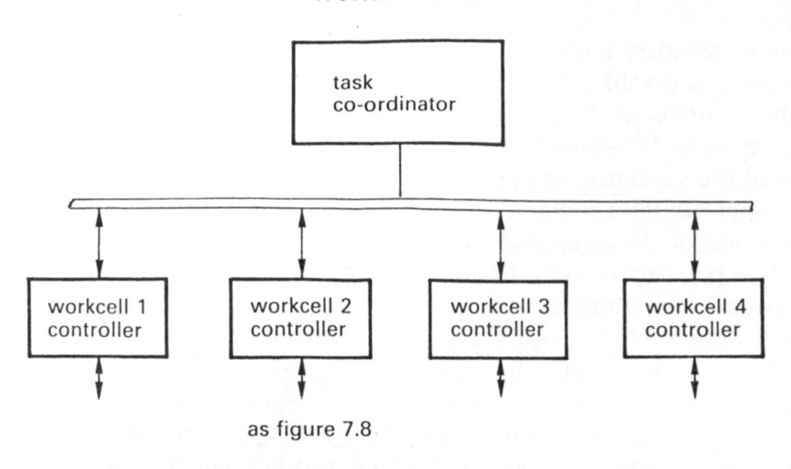


Figure 7.9 Co-ordinated cells

## 7.4 Collision avoidance

Clearly there should be no undesirable collisions between objects in the workcell. At the moment, this is ensured by the programmer (in the sense of the person who defines the various locations through which the robot will pass, the types of movement made and the logic of responses to sensory information) who will probably define the details of the task using his or her visualisation of the 3D world to estimate the safety of each action. This would be followed by a slow-speed execution of the task with the programmer's finger ready on the emergency stop button to halt the robot before an unforeseen collision occurs. If the program is faulty, then it must be edited and the above procedure repeated. Once the task has been verified to be correct at this stage, the speed would be increased up to the required value and the testing operation repeated.

The above manual procedure for collision avoidance is fine as long as the cell and task are relatively simple. If there are many branches in the program which are dependent on sensory information it can be very time consuming to perform thorough tests. This testing is performed on the target production robot system and is very expensive in terms of equipment utilisation. Two alternative approaches are used: online collision prediction through sensors, and offline simulation, prediction and path planning.

Online prediction of imminent collision is particularly prevalent in mobile robots, especially so when they must work in environments containing unpredictable humans in their working space. The various types of proximity sensors described in chapter 6 are used to detect when an object is dangerously close to the mobile robot which must be programmed to stop or take other evasive action.

Offline simulation software such as GRASP (Yong *et al.*, 1985) can be installed on a graphics workstation. The objects in the cell must be defined; details of commonly used devices, such as commercial robots, are automatically extracted from a database. The task itself must then be specified in terms of the sequences of operations, much as it would be programmed in the target runtime system. The simulation can then start, producing projections of 3D graphical images of the cell as the task proceeds. Different perspective views can be used and zoom facilities are provided so that critical operations can be studied in close-up. The use of this type of software also makes it easier to optimise the positions of devices in the workcell in order to maximise throughput. The user must, however, be aware of other considerations such as the repeatability and accuracy performance of the robot varying over its working volume. A limitation of such software, which applies to all simulators, is that the quality of the results will only be as good as the quality of the data and models put into it. If the fixtures are in practice slightly displaced from the positions specified at the simulation, or the robot is slightly out of calibration, then collisions can still occur on the target machine even though the simulation may indicate safe operation. It is wise, therefore, still to carry out the manual slow-speed check on the target cell, although only minor program changes should be necessary as a result.

## 7.5 Illustrative example: handling limp materials

The garment manufacturing industry, as will be discussed in the next chapter, is a potentially fruitful area for robotic applications. The particular problem (the motif application process) described here contains two features considered essential to future successful implementation of robotics in the industry. The first feature is the handling of limp materials, in this case the separation of a single panel of fabric from a stack of pieces, see figure 7.10. The separated piece must then be put in position on a worktable. The second feature is the use of low-cost vision techniques to cope with uncertainties in parts presentation, in this case exemplified by the need to extract a single motif, the correct way up, from a bin containing many such motifs. This is shown in figure 7.11. Following this, the motif must then be placed in the correct position and orientation on to the separated panel, giving the situation illustrated in figure 7.12 (with a different motif). The two items would then be fused together in a special press. For convenience, each of the two features and their solutions are described separately, followed by a description of the integration of the parts into a whole and a short note on further developments arising from this research.

Figure 7.10 A stack of cut fabric panels

Figure 7.11 A bin of motifs

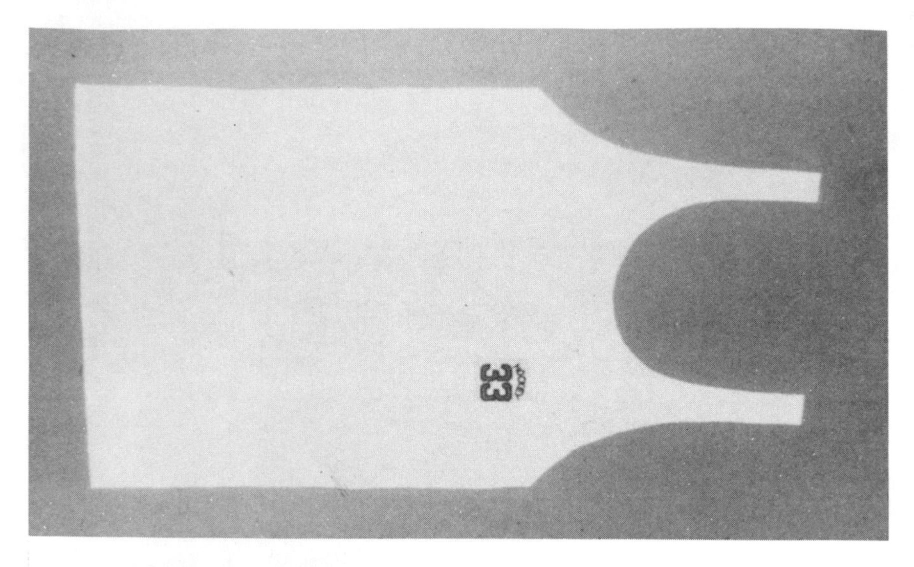

Figure 7.12 The final result — a motif on a separated panel

### *7.5.1 Fabric ply separation and placement task*

The main difficulty in separating a single ply of fabric from a stack such as that shown in figure 7.10, lies in the nature of the fabric. The primary requirement in this case was to handle knitted cotton fabric. This material has the properties of limpness, deformability varying according to direction, porosity, and the tendency for adjacent plies to cling together because of intertwining fibres, particularly at the cut edges. There are many potential ways of tackling such fabric handling problems (Taylor, P. M. and Koudis, 1987), but in this case a new method was invented (Kemp *et al.*, 1986) using a sensory air-jet/finger device shown schematically in figure 7.13 and attached to a PUMA robot. The principle of this device is that if an air-jet is directed at an angle over the fabric stack, it will induce vibrations in the top ply. In practice, because of the porosity of the fabric, some air penetrates the top ply, forming bubbles beneath it which are then propagated towards the edge of the stack. The vibrations and the moving bubbles break up the bonds between the plies, and do so particularly well at the edge. If the lower finger of the separation device of figure 7.13 is located just above the top edge of the stack, the vibrating top ply will flip over it, as seen in figure 7.14. This occurrence can be detected by an infra-red sensor in one finger which detects the amount of light received from the emitter located in the other finger (the cross-fire sensor of figure 7.13). A high level indicates no fabric between the jaws, a medium level indicates one piece only, and a low level indicates two or more plies. Once

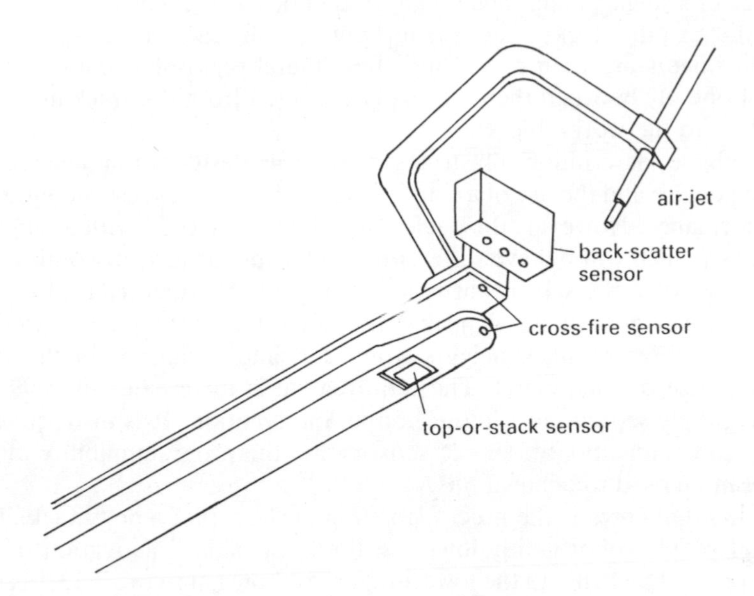


Figure 7.13 A schematic of an air-jet/finger ply separation device

Figure 7.14 Operation of the air-jet

the edge of a single ply has been separated in this way, the whole gripper is sliced through the stack, under the top ply, until it reaches the edge where the two fingers are then closed together, thereby gripping the complete edge of one ply between them. This ply is peeled from the stack and then placed on to the worktable.

The above description illustrates the basic design of a specialised gripping device and the use of a simple infra-red crossfire sensor to monitor its performance. However, the device must be able to cope with a range of materials (which will behave differently under the air-jet), discontinuities in the edge of the stack, changes in the top of the stack caused by the repeated removal of plies, changes in the initial top of the stack caused by the use of different stack heights when cut, and variations in the final gripping edge of each panel. The requirement is for greater than 99 per cent single ply separation without human intervention. It is instructive to see how a combination of simple sensors and the programmability of the robot can be used to achieve this.

The first unknown is the precise position of the top of a new stack. This is found by the robot gently lowering the gripper until activation of the switch protruding beneath the lower finger, as shown in figure 7.13, occurs. The gripper is then drawn across the top of the stack until the switch is deactivated, indicating that the edge has been reached. The location of this point is then entered automatically into a variable in the robot's program. The switch is connected to one of the robot's I/O lines which is interrogated after each increment of movement in the manner described in section 6.6.1.

The robot's controller must then determine which air-jet pressure is most suitable for the particular fabric being used. The air-jet is activated three times, with an increase in pressure each time. This is achieved via regulated air supplies switched using solenoids connected to three I/O lines. The signal from the reflective sensor (the back-scatter sensor of figure 7.13) at the end of the gripper is monitored during each activation of the air-jet. The amplitude of the periodic signal obtained from this sensor is related to the amplitude of the vibrations of the fabric, and therefore may be used to select the best pressure to use.

The gripper may now be moved to the first separation point, a fixed offset from the location variable determined above. The air-jet is switched on and the crossfire sensor interrogated after 0.5 seconds. If one piece is present between the fingers, the separation procedure continues as described above. If not, then the programmability of the robot allows multiple tries, a branch back to the find top and edge of stack routine, a modification of the gripper's separation position or, if all else fails, a pause with an error message so that an operator can intervene. This sensory capability raises the reliability of single ply separation from about 95 per cent to over 99 per cent.

### *7.5.2 Bin picking operation*

Motifs are made in a variety of shapes, sizes and colours. They are embroidered items with a backing of threads made from a nylon which has a low melting point. When subsequently heated and pressed against a fabric panel, the nylon melts and the motif fuses on to the fabric. It is crucial to put the motif the correct way up on the top of the fabric, otherwise it will fuse to the press instead of the fabric! The idea of a mechanical feeder was rejected because of the many different shapes of motif. Similarly, a vision system with a camera over the bin was not considered feasible because of the difficulty of analysing the images produced when the many different shapes and colours of motif are considered.

The approach taken was to use a blind vacuum pickup out of the bin, followed by the placement of whatever was picked up on to a rotation table under a camera. This simplifies the task for the vision system since the vacuum gripper would be most likely to pick up one motif only or at the most perhaps three motifs. Furthermore, the motifs will almost always lie flat on the table. The robot must then withdraw its gripper from the field of view of the camera so that a clear view of the motifs may be obtained.

The image processor must be triggered to grab an image and store it in memory. In this case, the standard video output of the cheap surveillance camera is sampled and thresholded by a special circuit to give a $64 \times 64$ binary image which is fed into the memory of the image processor via a parallel interface. The image processor then computes the area of the object within the field of view. Note that the lighting and threshold values (to obtain the binary image) must be correctly set to obtain a good quality image. These measures, and other thresholds referred to later, are automatically determined by the software in a teaching phase. Here, a perfect motif is placed on the table in the desired orientation and under good lighting conditions. Several images are grabbed in sequence and the thresholds of area are set to allow for noise in the image. This is repeated with the motif upside-down. In this case an object with a smaller area will be seen because of the whiteness of the protruding nylon threads. A secondary measure, the perimeter of the object, is also computed. The orientation procedure is invoked if both these measures fall within an acceptable tolerance band around the expected value obtained from a single motif which is the correct way up.

Orientation involves turning the rotation table in a series of steps. After each step an image is grabbed, centred in the software and template-matched against the taught image. The number of corresponding pixels in the two images which take differing values (pixel difference) is taken as the measure of mismatch between the images. Once the orientation is correct, then the two images should ideally be identical and this pixel difference measure will be zero. Experimental curves for one particular motif are shown in figure 7.15. The rotation continues in large steps of 20 degrees until the pixel difference falls below a threshold. Rotation then takes place in one degree steps until the minimum pixel difference is found. The motif is now correctly orientated, the location of its centroid is found, and it is now ready to be picked up and deposited on the separated fabric panel.

The program to handle the case of incorrect area or perimeter instructs the robot to take a particular action dependent on the information from the image processor. Similarly, the rotation table is also activated in response to information calculated by the same image processor. In this workcell, the same PUMA robot is used for both the ply separation and motif handling tasks. Clearly, there has to be communication between each part of the cell and co-ordination of their actions.

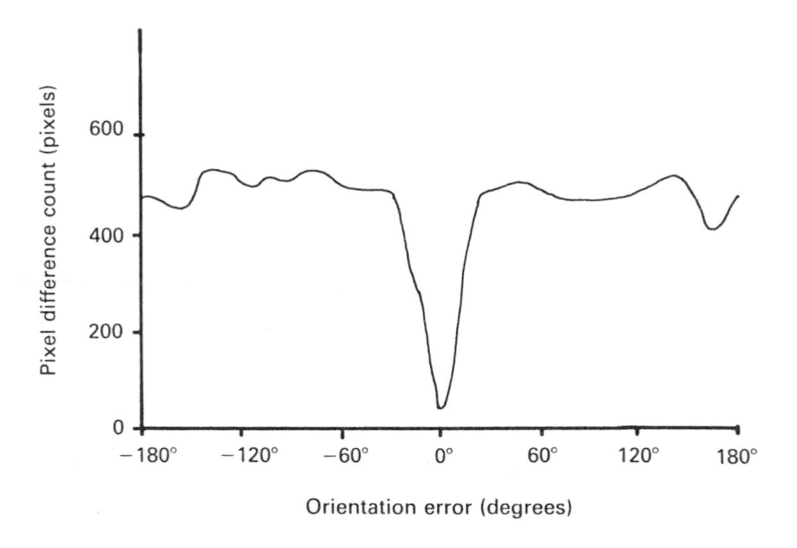


Figure 7.15 Template matching of motifs — experimental results

### *7.5.3 Integration*

Physical communication between devices is effected through an in-house bus called 'ROBUS'. Each peripheral, be it a robot, sensory processor or rotary table, is connected to the bus through a programmable interface card which buffers and passes messages to and fro with the correct protocols.

The above structure is only one of many possible options. In this case a major constraint was the use of version 1 of the VAL programming language which does not permit the complexity described above. It would, however, be possible to program the task description completely in version 2 of VAL, including communication to the rotation table and the LS111-based vision system. Similarly, it is also possible to squeeze both the controlling software and the image processing into a single LS111 processor which then also controls the rotation table and the PUMA.

As a postscript, it is worth noting that although the above system worked satisfactorily, it would be far too slow and expensive for industrial use. The main problems lie in the excessive cost and inadequate speed of the robot. However, the cell provided an excellent test bed for the development of the ply separation technique. Picking heads based on the air-jet/finger principle have been further developed and are commercially available. Similar vision feedback techniques are being used in continuing research projects towards flexible garment assembly.

# 8 Future Trends

Robotics still has a long way to go before it can break out of the quite restrictive industrial applications to which it is currently being applied. The majority of installations require little sensory capability or, indeed, flexibility of operation. The many reasons for this are best illustrated through a description of some of the current worldwide research work in robotics, and how this will affect the performance capabilities of future robotic installations. This leads naturally on to a discussion of future application areas for the technology. It should be borne in mind throughout this chapter that the relative cost of computing hardware is still falling. Therefore, projects which presently appear very expensive in computation time and hardware costs may well become viable in five or ten years time.

## 8.1 Robot arm design

Current robots are very heavy and most are slow compared with a human. Much of the current research work in arm design is concerned with lighter structures and different configurations.

It was emphasised in chapter 2 that it is important to keep heavy actuators as close to the base axes as possible. Rigidity of the links is required so that a stable end effector platform is provided, and so that the effector position may be determined from measurements of the joint angles. Rigidity is achieved by making the links suitably stiff and therefore heavy. Reducing the weight and hence the static stiffness of the arm means that the required stiffness must be supplied via the sensing and control of the deflections. This is akin to the way an angler can precisely place a very light and flexible fishing rod.

One of the main difficulties lies not so much in the compensation for static deflections because of loading, but more in the suppression of unwanted oscillations at the end of the arm's travel. A flexible arm will have several natural modes of vibration. It is possible to compensate for these oscillatory modes only if the actuators have a sufficiently high bandwidth. Thus the more flexible the arm, the lower are its natural frequencies of vibration and the easier it is to compensate for them. Unfortunately, higher harmonics may also be significant. In addition, the arm may have forces applied to it in more than one sense; the shoulder-to-elbow link of the SCARA type mechanism of figure 1.4 will see vertical, horizontal and torsional force applied to it during movement. Depending on the link's construction, there may be significant modes of vibration about all three axes. Thus it may still be necessary to make a link stiff about one or more axes. The actuators may then be controlled to compensate for the remaining modes of flexibility.

The analysis of flexible link arms is even more complex than that for the rigid link arms considered in chapter 2. Nevertheless, encouraging results are being obtained (Truckenbrodt, 1981; Fresonke *et al.*, 1988). It is likely that the first land-based applications will be in very large robots for the nuclear industry.

Another future trend is likely to be the greater use of modular axes which can be configured at will by the user to form a distributed robotic system. A suitable combination of simple axes may yield a faster and cheaper result than that achieved by use of a sophisticated robot where most of its capabilities may be under-utilised for much of the working cycle.

## 8.2 Actuators

The improvement of conventional electric motors has been gradual. However, the likely emergence of superconducting materials which operate at room temperatures could lead to important breakthroughs in motor performance. These motors would have negligible resistive heat losses and could therefore be operated with much higher currents. Linear actuators could also be made.

The use of water as a hydraulic fluid is also under study, particularly for hazardous and undersea environments.

Various polymers exhibit electrorestriction, that is they contract or expand when an electrical field is applied to them. Other polymers exhibit magnetorestriction. Such materials can provide small movements with very high forces and low power losses. Finally, some co-polymers expand and contract when various chemicals are applied to them. A co-polymer such as Polyvinyl alcohol combined with Polyacrylic acid (PVA/PAA) can exhibit strain rates of 10 per cent/s and total strains of over 100 per cent before breakage.

## 8.3 Sensors and sensory data processing

There are few vision sensors designed specifically for robotic applications. As noted in chapter 6, most vision sensors have been developed primarily for the television and home video industries. The mass markets in these areas ensure relatively low cost devices but their performance specification is not always satisfactory. One of the major restrictions is the output of a video signal at the standard TV frame rates. A robot's end effector moving at a modest 1 metre/s will have travelled 4 cm while an interlaced image is built up. A greater light sensitivity and faster sampling rates are needed if such a camera is to be used for real-time tracking. Solid state vision sensing devices are becoming more prominent and have the advantages of lightness and robustness. It may also be that in the near future it will be possible to purchase such devices with a selection of on-chip simple image processing facilities. Thus, rather than having to grab a complete image from the camera, transfer it to the memory of a computer and then process this data to, say, find the centroid of a object, one could merely send a signal to the camera/processor and have the centroid co-ordinates returned.

The traditional rectangular image may not be the most suitable for robotic use. An interesting alternative has been proposed (Inigo *et al.*, 1986) which has similarities to the human visual system. The pixels are now arranged in circular form, getting smaller close to the centre of the circle and larger away from it, see figure 8.1. So far this option has only been simulated.

Fibre optic sensors comprise one or more optical fibres with specially treated ends. In one simple example, if a laser is shone down an optical fibre which has a reflective end, then the light will be reflected back to the source. If a beam splitter is used, the phase between the source and the

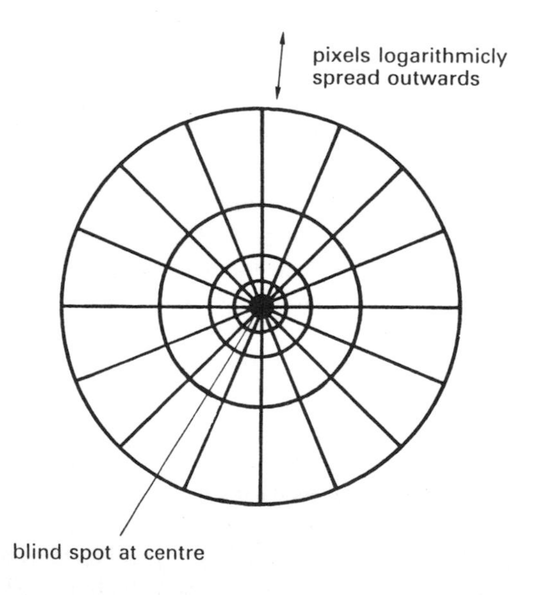


Figure 8.1 Circular form of pixel topology

reflected waveforms can be compared (Jones and Jackson, 1986). If the length of the fibre changes, then so will the phase of the reflected waveform change, since it has more (or less) distance to travel. We have, therefore, an optical version of a strain gauge. Other coatings can be applied to the end of the fibre, making it sensitive to certain chemicals. This, and other chemically sensitive semiconductors, could be used to give robots a sense of smell.

Parallel topology computers can perform many image processing operations much faster than a serial digital computer. Researchers have worked for several years with specialised parallel computers such as CLIP (Duff, 1982). These have been relatively expensive but the advent of the Transputer (Inmos, 1986) augurs well for a low-cost alternative with superior performance.

A different approach to pattern recognition is exemplified by WISARD (Wilkie, Stonham and Aleksander's Recognition Device) which is intended to be a simplified facsimile of recognition functions of the brain. The key feature of this device is the way that the address terminals of the RAMs which hold the image are connected up at random. A most readable account of this device, its place within artificial intelligence, and possible developments, is given by Aleksander and Burnett (1983).

## 8.4 Dynamic control

This is a very active area of research as control engineers continue to devise ways of obtaining better dynamic performance from robots. The difficulties of obtaining good control arise from the interactions between joints, the non-linear nature of the dynamical equations, variations in the load being carried, and non-linearities in the actuators and drive trains. A further constraint is that the control has to be performed in real time, placing a large burden on the computational hardware. However, computational costs are decreasing and developments such as the Transputer offer a potential for very fast parallel computing to be harnessed to these problems.

In chapter 4 the dynamic control of a multilink robot was described, with each axis controlled individually. The interactive effects were compensated for by the use of feedforward terms. The problem with feedforward comes when the interactive terms are not what was expected. Incorrect compensation is then applied and the performance degrades accordingly. Multivariable control system design, using techniques such as those found in the survey of MacFarlane (1979), aims for the design of an interactive compensator which will try to fully, or more usually partially, decouple each axis (Williams, 1985).

Adaptive controllers use variation of the compensator parameters to try to give a constant dynamic performance even as the plant dynamics change. Improved performance can be obtained if the controller can anticipate the following demand position (Zarrop, 1987).

Non-linear controllers give the designer more freedom and can lead to time-optimal, or near time-optimal, responses, and as such have been proposed for robotic applications (Young, 1978). Figures 8.2 show a *variable structure* type of controller applied to a single axis of the UMI-RTX robot. This robot has significant non-linearities in its drive train and has been modelled as shown in figure 8.3. The basic controller is quite simple, effecting switching between negative and positive feedback according to the values of positional error and output velocity. Refinements are added to reduce steady state errors and smooth the switching action. Further details of the design of this controller are given by Taylor, P. M. and Gilbert (1987) who have since extended the application to the control of three axes simultaneously.

An interesting approach has been devised by Hewit (1983) and is depicted in figure 8.4. The technique requires measurement of the accelerations of the links. An idealised approximation of the theoretical inertia matrix $I(\boldsymbol{\theta})$ is multiplied with the above-measured acceleration vector (after conversion to world co-ordinates) to give a force vector. This vector contains those forces which, if applied to the theoretical arm, would give the actual measured accelerations. The difference between these forces and the measured vector of actual applied forces corresponds to 'noise'. This arises from signal noise, measurement errors and errors in the model because of simplifications such as neglect of non-linearities. The principle of this type of control is that the 'noise' signal is negated and fed into the motor drivers such that the true 'noise' is cancelled out. The resulting system should then act as if it actually had the simplified inertia matrix $I(\boldsymbol{\theta})$

As alluded to above, the control of a flexible arm is considerably more difficult than that of a rigid arm. Bayo (1988) describes some progress being made in this area.

## 8.5 Workcell integration and sensory task control

There are many effective ways of integrating the various parts of a workcell. It is really a standardisation issue rather than one where there are major technical or theoretical problems to be overcome. Performance and cheapness are major factors. These comments also apply to the task of integrating numbers of cells into a factory automation system.

Many components are now designed using Computer Aided Design Workstations. In such cases, there is much information about the compo-

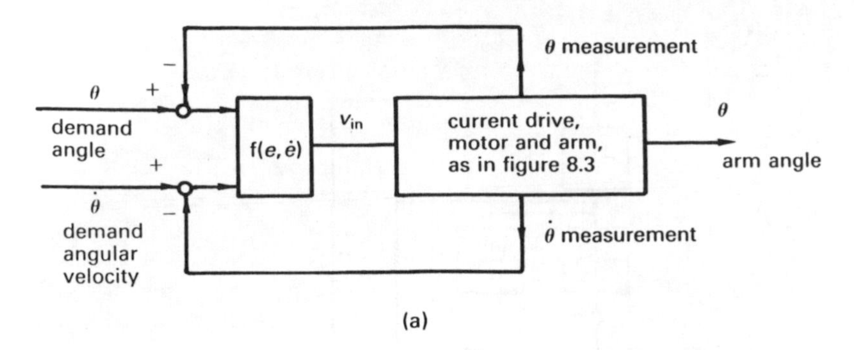


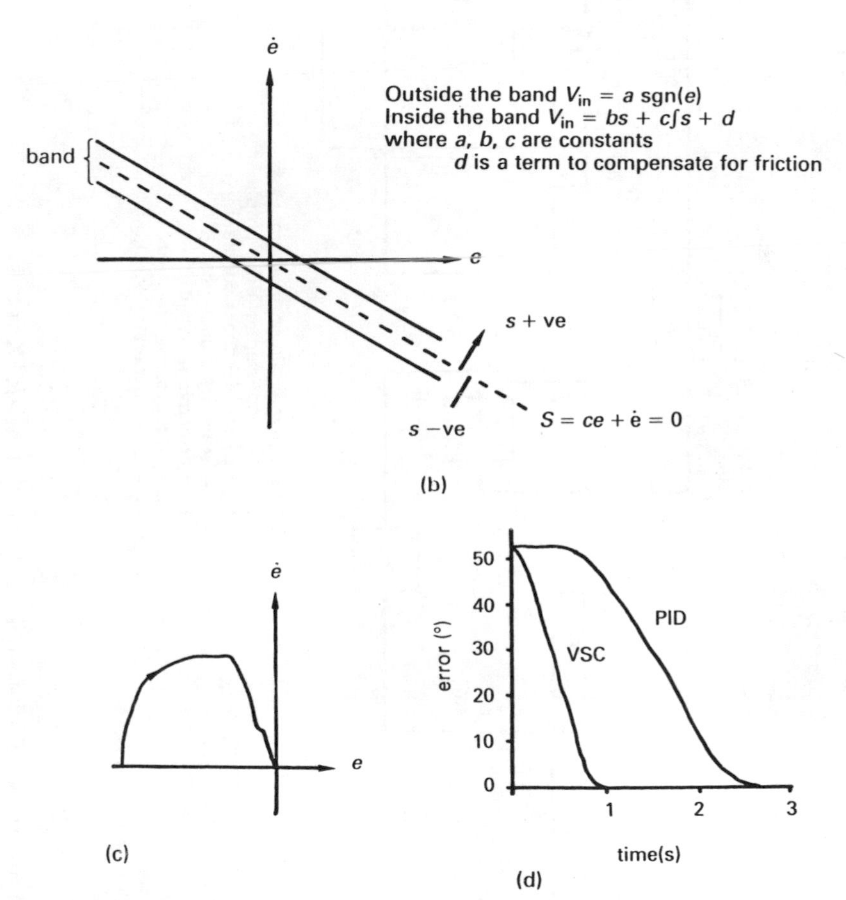


Figure 8.2 Variable structure control applied to a single axis of the UMI-RTX robot: (a) control structure; (b) switching controller of $(e, \dot{e})$; (c) a typical experimental response in the $(e, \dot{e})$ phase plane; (d) comparison of step responses with a PID control and variable structure control

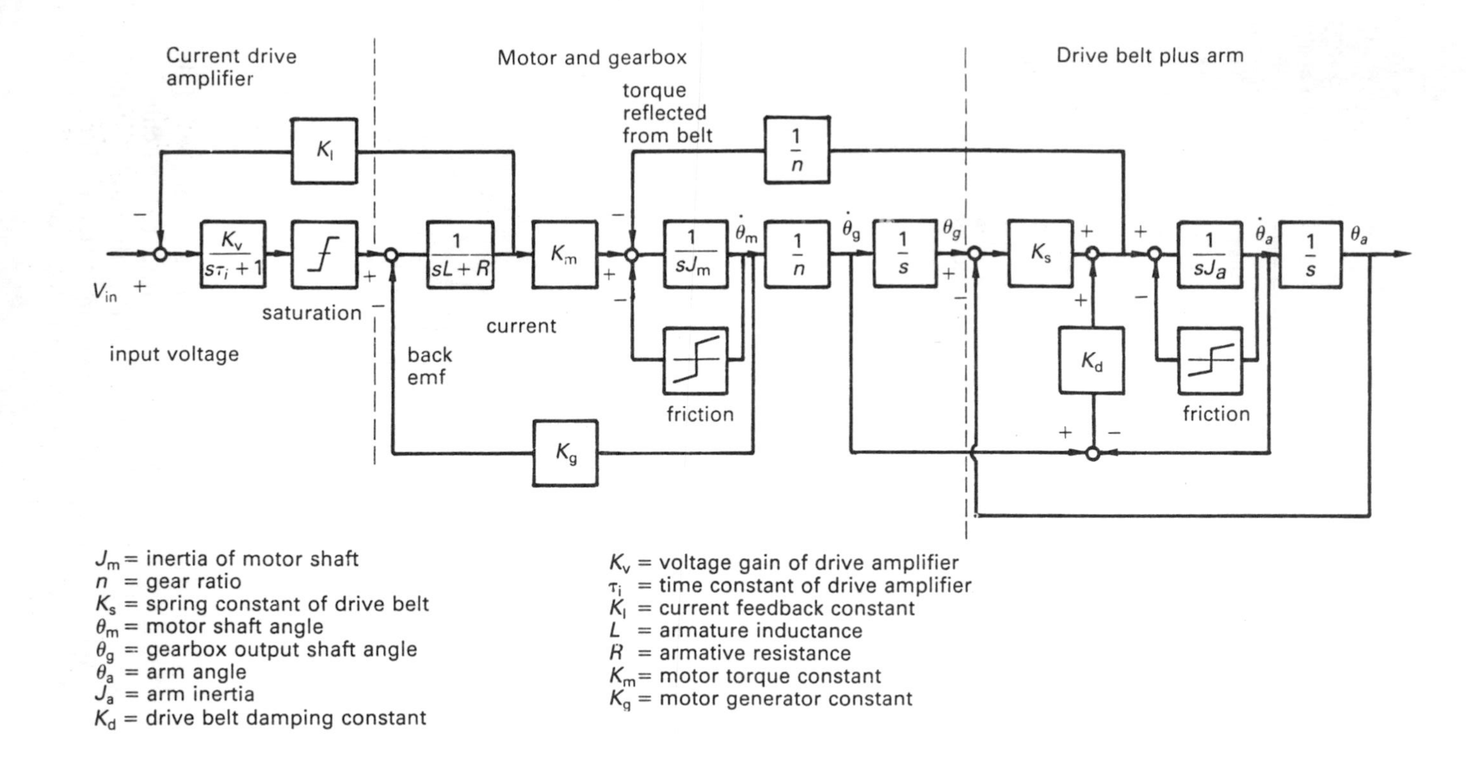


Figure 8.3 Non-linear model of a single axis of the UMI-RTX robot

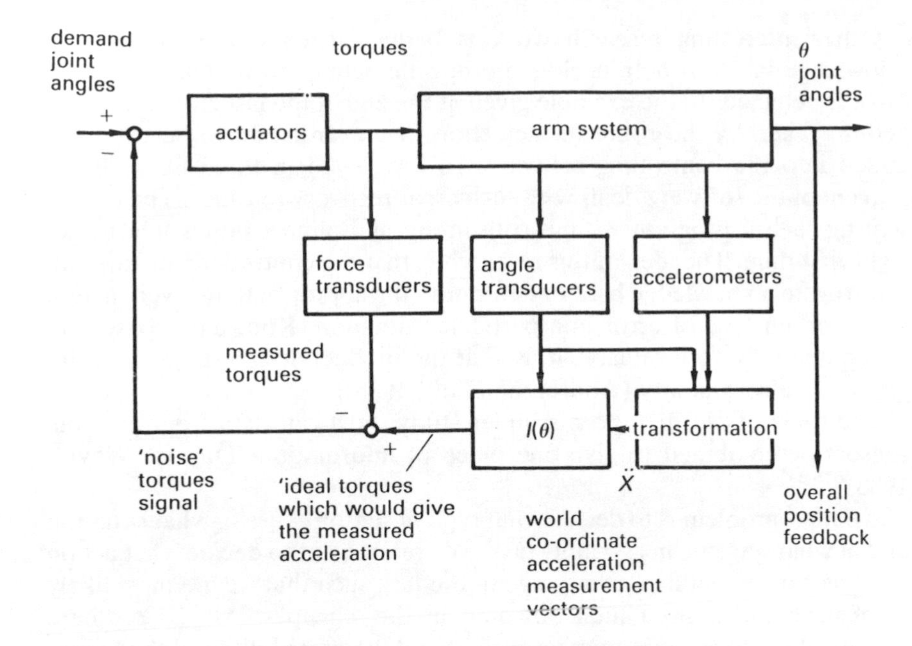


Figure 8.4 'Invariant control' scheme of Hewit (1983). Overall feedback control and dynamic compensation not shown

nents which could be valuable during the planning of a robot task. Let us take automated assembly as an example. It is technically quite feasible for the designer not only to define the components used in the assembly, but also to specify their geometrical relationships with each other once assembled. If we ignore any sensing capability at the moment, then the component feeder positions could be defined and a program executed to plan the assembly, working out the sequence of movements required. This assembly plan could be translated into whatever language is used for the workcell control. As an intermediate stage, it would be quite useful if the assembly task could be described at the object level for subsequent translation into manipulator level commands for the workcell controller. Such an object level language, RAPT, has been developed at the University of Edinburgh (Ambler *et al.*, 1975), and an assembly task can be automatically translated into VAL for a PUMA robot (Ambler *et al.*, 1982).

Some progress has also been made (Yin *et al.*, 1984) in allowing vision verification, in which variables are associated with some locations and values inserted into them as a result of runtime sensory information. Once vision has been used to find, say, one edge of an object A, the world model is updated, and an inference program is run to change the locations of all other objects connected to object A.

Other interesting research work is being carried out in the use of knowledge bases to help decide appropriate actions to be taken once an error is detected. In the example given at the end of the last chapter, all the actions taken by the robot on detection of an error had to be explicitly coded into the controlling software. In fact, even in this case, over 90 percent of the software deals with such error recovery routines. The task of writing such a program is time consuming and almost impossible to get right first time. The idea of this research is that the controller software can interrogate a knowledge base to determine the appropriate recovery action for a certain type of error in a particular situation. Knowledge bases are also potentially very valuable if used at the product design stage to ensure good manufacturability (Andreasen *et al.*, 1988).

The topic of *sensor fusion* is under study: how can data from multiple sensors be combined to give one piece of information (Durrant-Whyte, 1988)?

Another problem is to decide what type of sensor to use in what situation and at what stage in an assembly task. We really want to decide what action must be carried out at a given stage in the task such that we are most likely to achieve the desired final outcome in the 'cheapest' way. Cost here should take into account time resources, and the probabilities of the failure of each action due to faults in components, actuators, fixtures and inherent flaws in the particular operation being carried out (Taylor, P. M. 1988). Such decision making will require extensive knowledge bases to be set up. These will contain generic information such as knowledge of peg-in-hole operations plus detailed knowledge of the likely performance of cell elements. It is possible that this knowledge base might be updated according to the runtime experiences of the cell.

Such control problems are also generic to the control of large dynamic systems. So-called expert systems essentially produce a set of if–then rules which specify the actions to take as a result of sensor signals being true or false. This scheme has a limited learning capability. Other workers have proposed the use of fuzzy set theory which can be used to characterise information which is known in a qualitative rather than quantitative sense. A third approach put forward by other researchers is based on the ideas of neural networks, similar to the ideas behind WISARD, mentioned in section 8.2. A readable introduction to neural networks is given by Bavarian (1988) in a special section of the *IEEE Control Systems Magazine*, where further more detailed papers may be found.

At a higher level lies the design of the overall task itself. There will be many possible ways of assembling a particular product. If estimates of the cost of a particular way can be found, as outlined above, then it should be possible to select the one which is likely to be the 'cheapest'. Indeed there may be many ways of designing a specific product. Consideration of likely assembly problems at this stage, so-called *design for assembly* (Andreasen *et al.*, 1988) can appreciably simplify the implementation and operation of the assembly workcell.

## 8.6 Application areas

Robotics is not the instant answer to all industrialists' prayers. Many problems are best solved by designing them away in the first place, rather than by throwing technology at them. However, this is not always a possible option. In addition, cost-effective sensing will enable robotic techniques to be applied in many new areas. This is particularly so where any form of natural material is being used, so that the robotic system can cope with the inevitable variations in the products being handled. In cases such as fabric handling, the materials are limp and sometimes behave in a 'live' manner; single knitted cotton material, for example, tends to curl up at the edges when repeatedly handled. There is, therefore, potential for robotic applications in garment manufacture, shoe manufacture, and agriculture in terms of plant and animal husbandry. The classic case of animal handling is a laboratory prototype sheep-shearing robot (Wong and Hudson, 1983) which combines sensing with rapid actuation of the end effector to ensure close shearing without damage to the animal. Other researchers are working on topics such as automated milking of cows (Montalescot, 1986). Fruit picking is being studied throughout the world with experimental work on orange harvesting (Coppock, 1984) and apple picking (Grand d'Enson, 1985). Transplanting machines (Krutz *et al.*, 1986) and vine pruning (Monsion *et al.*, 1985; Peyran *et al.*, 1986) are also being studied. Baylou (1987) gives an excellent survey of robotics in the agricultural area.

There is much potential for the use of robotic devices in hazardous environments, such as fire fighting, bomb disposal and working on high buildings (such as cleaning). Eventually, mobile robots could be developed for household work , with perhaps the first application being for autonomous cleaning devices (Yasutoru *et al.*, 1988).

# Appendix A: Relationships between Co-ordinate Frames

Let us consider a rectangular co-ordinate frame, frame 0, fixed to the ground, and a rectangular co-ordinate frame, frame 1, fixed to an object at an arbitrary position in space. In general, the origins of the two frames will not be coincident and the corresponding unit vectors (the *basis vectors*) making up each co-ordinate system will not be parallel.

Consider the case shown in figure A.1 where the corresponding basis vectors in each frame are parallel and we may say that the frames have the same orientation. The origins are displaced such that the origin of frame 1 is at $\boldsymbol{p} = [p_x, p_y, p_z]^{\mathrm{T}}$ with respect to the origin of frame 0. The relationship

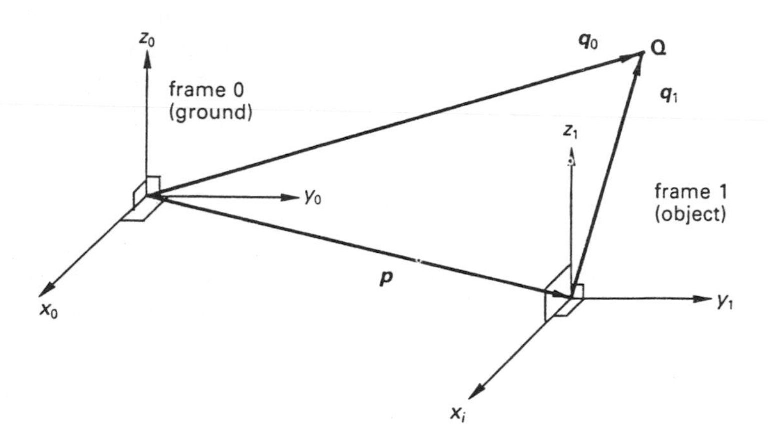


Figure A.1

between a vector $\boldsymbol{q}_0$ and the vector $\boldsymbol{q}_1$ defining a point Q in co-ordinate frames 0 and 1 respectively is clearly

$$\boldsymbol{q}_0 = \boldsymbol{q}_1 + \boldsymbol{p}$$

Consider now the case where the origins are co-incident but the frames no longer have the same orientation. In figure A.2 the difference in orientation corresponds to a rotation about the $z$-axis.

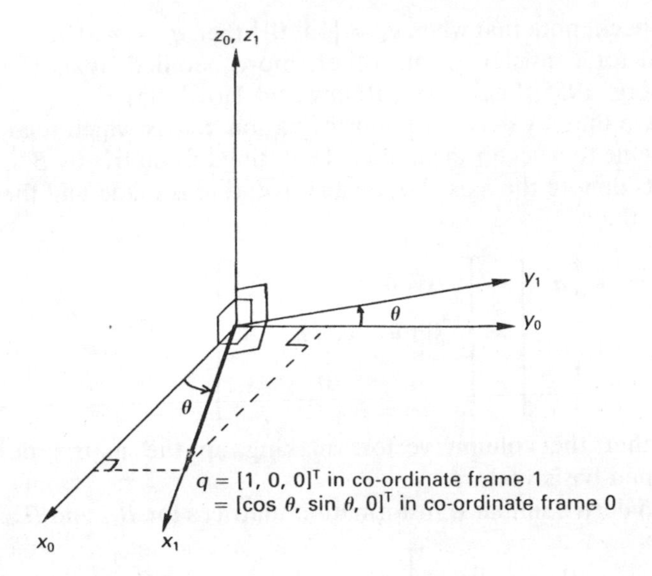


Figure A.2

Consider a unit vector $\boldsymbol{q}$ along $x_1$, that is, $\boldsymbol{q} = [1,0,0]^T$. Then by inspection of figure A.2, if this vector is expressed as $\boldsymbol{n}$ in co-ordinate frame 0, then

$$\boldsymbol{n} = [\cos\theta, \sin\theta, 0]^T$$

Similarly, a unit vector along $y_1$, $\boldsymbol{q} = [0,1,0]^T$ may be expressed in co-ordinate frame 0 as

$$\boldsymbol{o} = [-\sin\theta, \cos\theta, 0]^T$$

The unit vector $\boldsymbol{q}$ along $z_1$ is unchanged and therefore, when transformed into co-ordinate frame 0, becomes

$$\boldsymbol{a} = [0,0,1]^T$$

By superposition, we may now find the representation $\boldsymbol{q}_0$ in co-ordinate frame 0 of any vector $\boldsymbol{q}_1$ in co-ordinate frame 1 when frame 1 is rotated by an angle θ, relative to frame 0, about the $z$-axes. Thus

$$\boldsymbol{q}_0 = \left[\begin{array}{c:c:c} \boldsymbol{n} & \boldsymbol{o} & \boldsymbol{a} \\ & & \\ & & \end{array}\right] \boldsymbol{q}_1$$

As a check, note that when $\boldsymbol{q}_1 = [1,0,0]^T$ then $\boldsymbol{q}_0 = \boldsymbol{n}$. The notation $\boldsymbol{n}$, $\boldsymbol{o}$, $\boldsymbol{a}$ is used for consistency with other, more detailed, texts (Denavit and Hartenberg, 1955; Paul, 1981; Ranky and Ho, 1985).

We have thereby derived a transformation matrix which relates a vector in one frame to a vector in another. Denoting this matrix by $B_{z\theta}$, where the subscripts denote the axis about which rotation is made and the degree of rotation, then

$$B_{z\theta} = \left[\begin{array}{c|c|c} \boldsymbol{n} & \boldsymbol{o} & \boldsymbol{a} \\ & & \\ & & \end{array}\right] = \begin{bmatrix} \cos\theta & -\sin\theta & 0 \\ \sin\theta & \cos\theta & 0 \\ 0 & 0 & 1 \end{bmatrix}$$

Note that the column vectors making up the matrix describe the transformed basis vectors.

We can derive similar transformation matrices for $B_{x\theta}$ and $B_{y\theta}$. They are

$$B_{x\theta} = \begin{bmatrix} 1 & 0 & 0 \\ 0 & \cos\theta & -\sin\theta \\ 0 & \sin\theta & \cos\theta \end{bmatrix}$$

$$B_{y\theta} = \begin{bmatrix} \cos\theta & 0 & \sin\theta \\ 0 & 1 & 0 \\ -\sin\theta & 0 & \cos\theta \end{bmatrix}$$

In order to consider any general orientation, we need to analyse the results of a series of rotations about a sequence of axes. Thus, in terms of the Euler angles (see section 2.1), frame 1 might be obtained from frame 0 by the sequence: rotate frame 0 by $\phi$ about its $z$-axis to form frame H, rotate frame H by $\theta$ about its $y$-axis to form frame I, rotate frame I by $\psi$ about its $z$-axis to form frame 0. Thus

$$\boldsymbol{q}_0 = B_{z\phi}\boldsymbol{q}_H, \quad \boldsymbol{q}_H = B_{y\theta}\boldsymbol{q}_I, \quad \boldsymbol{q}_I = B_{z\psi}\boldsymbol{q}_1$$

Combining these transformations gives

$$\boldsymbol{q}_0 = B_{z\phi}B_{y\theta}B_{z\psi}\boldsymbol{q}_1$$

thus giving a transformation matrix $B_{\text{Euler }\phi\theta\psi}$ given by

$$B_{\text{Euler }\phi\theta\psi} = B_{z\phi}B_{y\theta}B_{z\psi}$$

$$= \begin{bmatrix} \cos\phi\cos\theta\cos\psi - \sin\phi\sin\psi & -\cos\phi\cos\theta\sin\psi - \sin\phi\cos\psi & \cos\phi\sin\theta \\ \sin\phi\cos\theta\cos\psi + \cos\phi\sin\psi & -\sin\phi\cos\theta\sin\psi + \cos\phi\cos\psi & \sin\phi\sin\theta \\ -\sin\theta\cos\psi & \sin\theta\sin\psi & \cos\theta \end{bmatrix}$$

Similarly, a transformation matrix $B_{RPY}$ may be derived for roll, pitch and yaw, as

$$B_{RPY} = \begin{bmatrix} \cos\phi\cos\theta & \cos\phi\sin\theta\sin\psi - \sin\phi\cos\psi & \cos\phi\sin\theta\cos\psi + \sin\phi\sin\psi \\ \sin\phi\cos\theta & \sin\phi\sin\theta\sin\psi + \cos\phi\cos\psi & \sin\phi\sin\theta\cos\psi - \cos\phi\sin\psi \\ -\sin\theta & \cos\theta\sin\psi & \cos\theta\cos\psi \end{bmatrix}$$

Given a specified transformation matrix $B$ it is possible, as is shown in section 2.5.4 and appendix B, to obtain the equivalent Euler angles. Therefore, the matrix $B$ gives complete information on the orientation of one frame with respect to another.

The next step is to invoke superposition again to see how a vector $\boldsymbol{q}_1$ in frame 1 is transformed into vector $\boldsymbol{q}_0$ in frame 0 when the two frames are mis-aligned *and* the origins are translated. Thus

$$\boldsymbol{q} = B\boldsymbol{q}_1 + \boldsymbol{p} \tag{A.1}$$

where $B$ is the appropriate orientation transformation as described above and $\boldsymbol{p}$ is the translation between origins expressed in co-ordinate frame 0.

There are many instances in robotics where different co-ordinate frames are used. In chapter 2 and appendix B, co-ordinate frames are fixed to each link of the robot. Co-ordinate frames may also be fixed to cameras and other sensors mounted at various locations in the workplace. It is essential that information expressed in one frame can be translated into information in another frame. The transformation matrix $B$ and the translation vector $\boldsymbol{p}$ allow us to do this.

# Appendix B: Joint Transformations and their Application to an Elbow Manipulator

Links $i$ and $(i - 1)$ and revolute joints $(i - 1)$, $i$ and $(i + 1)$ of an arbitrary manipulator are shown in figure B.1.

The $i$th co-ordinate frame is attached to the $i$th link at joint $(i + 1)$. It is convenient for the $z$-axis to lie along the axis of motion of the $(i + 1)$th joint. The $x_i$-axis is normal to the $z_{i-1}$-axis and points away from it. The angle $\alpha_i$ is called the *twist angle* of link $i$.

The $i$th frame may be obtained from the $(i - 1)$th frame by the following sequence of transformations:

1. Rotate by an angle $\theta_i$ about $z_{i-1}$.
2. Translate along $z_{i-1}$ a distance $d_i$ (note that $x_{i-1}$ has now been rotated to be parallel to $x_i$).
3. Rotate about $x_i$ by the twist angle $\alpha_i$.

Thus symbolically

$$A_{i-1,i} = \mathrm{ROT}(z_{i-1}, \theta)\mathrm{TRANS}(0, 0, d_i)\mathrm{TRANS}(\alpha_i, 0, 0)\mathrm{ROT}(x_i, \alpha_i)$$

where $\mathrm{ROT}(z, \theta)$ corresponds to an augmented rotation transformation $B_{z\theta}$, and $\mathrm{TRANS}(a, b, c)$ corresponds to an augmented translation transformation $\boldsymbol{p} = [a, b, c]^{\mathrm{T}}$.

By substitution of the appropriate transformation matrices we get

$$A_{i-1,i} = \begin{bmatrix} \cos\theta_i & -\sin\theta_i & 0 & 0 \\ \sin\theta_i & \cos\theta_i & 0 & 0 \\ 0 & 0 & 1 & 0 \\ 0 & 0 & 0 & 1 \end{bmatrix} \begin{bmatrix} 1 & 0 & 0 & a_i \\ 0 & 1 & 0 & 0 \\ 0 & 0 & 1 & d_i \\ 0 & 0 & 0 & 1 \end{bmatrix} \begin{bmatrix} 1 & 0 & 0 & 0 \\ 0 & \cos\alpha_i & -\sin\alpha_i & 0 \\ 0 & \sin\alpha_i & \cos\alpha_i & 0 \\ 0 & 0 & 0 & 1 \end{bmatrix}$$

which on multiplying out gives

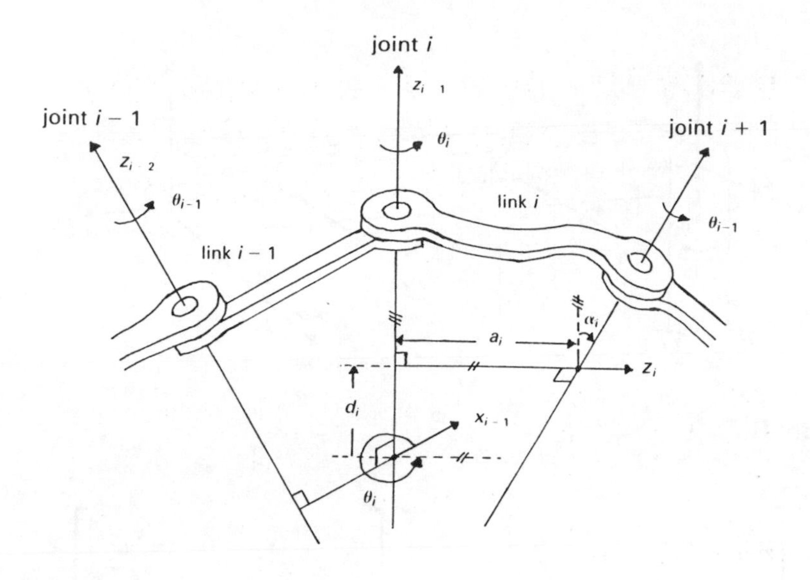


Figure B.1

$$A_{i-1,\, i} = \begin{bmatrix} \cos\theta_i & -\cos\alpha_i \sin\theta_i & \sin\alpha_i \sin\theta_i & a_i \cos\theta_i \\ \sin\theta_i & \cos\alpha_i \cos\theta_i & -\sin\alpha_i \cos\theta_i & a_i \sin\theta_i \\ 0 & \sin\alpha_i & \cos\alpha_i & d_i \\ 0 & 0 & 0 & 1 \end{bmatrix}$$

For a prismatic joint, the $A$ matrix becomes (Paul, 1981):

$$A_{i-1,\, i} = \begin{bmatrix} \cos\theta_i & -\cos\alpha_i \sin\theta_i & \sin\alpha_i \sin\theta_i & 0 \\ \sin\theta_i & \cos\alpha_i \cos\theta_i & -\sin\alpha_i \cos\theta_i & 0 \\ 0 & \sin\alpha_i & \cos\alpha_i & d_i \\ 0 & 0 & 0 & 1 \end{bmatrix}$$

If the above techniques are applied to the elbow manipulator shown in figure B.2, the transformations become (Paul, 1981):

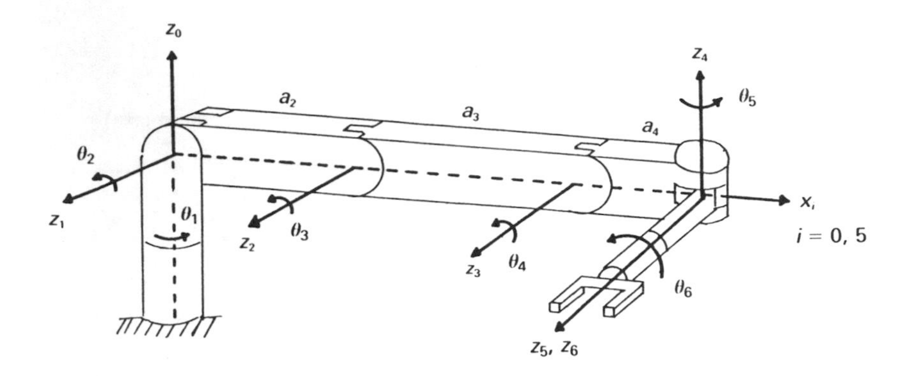


Figure B.2

$$A_{01} = \begin{bmatrix} C_1 & 0 & S_1 & 0 \\ S_1 & 0 & -C_1 & 0 \\ 0 & 1 & 0 & 0 \\ 0 & 0 & 0 & 1 \end{bmatrix} \quad A_{12} = \begin{bmatrix} C_2 & -S_2 & 0 & C_2a_2 \\ S_2 & C_2 & 0 & S_2a_2 \\ 0 & 0 & 1 & 0 \\ 0 & 0 & 0 & 1 \end{bmatrix}$$

$$A_{23} = \begin{bmatrix} C_3 & -S_3 & 0 & C_3a_3 \\ S_3 & C_3 & 0 & S_3a_3 \\ 0 & 0 & 1 & 0 \\ 0 & 0 & 0 & 1 \end{bmatrix} \quad A_{34} = \begin{bmatrix} C_4 & 0 & -S_4 & C_4a_4 \\ S_4 & 0 & C_4 & S_4a_4 \\ 0 & -1 & 0 & 0 \\ 0 & 0 & 0 & 1 \end{bmatrix}$$

$$A_{45} = \begin{bmatrix} C_5 & 0 & S_5 & 0 \\ S_5 & 0 & -C_5 & 0 \\ 0 & 1 & 0 & 0 \\ 0 & 0 & 0 & 1 \end{bmatrix} \quad A_{56} = \begin{bmatrix} C_6 & -S_6 & 0 & 0 \\ S_6 & C_6 & 0 & 0 \\ 0 & 0 & 1 & 0 \\ 0 & 0 & 0 & 1 \end{bmatrix}$$

where $C_i$ indicates $\cos \theta_i$ and $S_i$ denotes $\sin \theta_i$.

The product of these $T = A_{01}\, A_{12}\, A_{23}\, A_{34}\, A_{45}\, A_{56}$ is of the standard form

$$\boldsymbol{T} = \left[\begin{array}{c|c|c|c} \boldsymbol{n} & \boldsymbol{o} & \boldsymbol{a} & \boldsymbol{p} \\ & & & \\ \hline 0 & 0 & 0 & 1 \end{array}\right]$$

where

$$\boldsymbol{n} = \begin{bmatrix} C_1(C_{234}C_5C_6 - S_{234}S_6) - S_1S_5C_6 \\ S_1(C_{234}C_5C_6 - S_{234}S_6) + C_1S_5C_6 \\ S_{234}C_5C_6 + C_{234}S_6 \end{bmatrix}$$

$$\boldsymbol{o} = \begin{bmatrix} -C_1(C_{234}C_5S_6 + S_{234}C_6) + S_1S_5S_6 \\ -S_1(C_{234}C_5S_6 + S_{234}C_6) - C_1S_5S_6 \\ -S_{234}C_5S_6 + C_{234}C_6 \end{bmatrix}$$

$$\boldsymbol{a} = \begin{bmatrix} C_1C_{234}S_5 + S_1C_5 \\ S_1C_{234}S_5 - C_1C_5 \\ S_{234}S_5 \end{bmatrix}$$

$$\boldsymbol{p} = \begin{bmatrix} C_1(C_{234}a_4 + C_{23}a_3 + C_2a_2) \\ S_1(C_{234}a_4 + C_{23}a_3 + C_2a_2) \\ S_{234}a_4 + S_{23}a_3 + S_2a_2 \end{bmatrix}$$

and $C_{ijk}$ indicates $\cos(\theta_i + \theta_j + \theta_k)$.

The above results let us find the transformation matrix $T$ given the values of the joint angles $\theta_i$. Sometimes we will want to find the values of the joint angles given the values of the elements of $T$, for example, when wanting to move the robot to a position specified in world co-ordinates (thereby defining $T$).

Considering the elbow manipulator of figure B.2, it can be shown (Paul, 1981, pp. 78–82) that the joint angles $\theta_i$ are given by:

$$\theta_1 = \tan^{-1}(p_y/p_x) \text{ or } \theta_1 = 180° + \tan^{-1}(p_y/p_x)$$

$$\theta_2 = \tan^{-1}\left(\frac{(C_3a_3 + a_2)p'_y - S_3a_3p'_x}{(C_3a_3 + a_2)p'_x + S_3a_3p'_y}\right)$$

where $p'_x = C_1p'_x + S_1p'_y - C_{234}a_4$

and $p'_y = p_3 - S_{234}a_4$

and $\theta_{234} = \tan^{-1}\left(\frac{a_z}{C_1a_x + S_1a_y}\right)$ or $\theta_{234} = \theta_{234} + 180°$

and $C_{ijk} = \cos(\theta_i + \theta_j + \theta_k)$ and $S_{ijk} = \sin(\theta_i + \theta_j + \theta_k)$

$$\theta_3 = \cos^{-1}\left(\frac{p'^2_x + p'^2_y - a_3^2 - a_2^2}{2a_2a_3}\right)$$

$$\theta_4 = \theta_{234} - \theta_3 - \theta_2$$

$$\theta_5 = \tan^{-1}\left(\frac{C_{234}(C_1a_x + S_1a_y) + S_{234}a_z}{S_1a_x - C_1a_y}\right)$$

$$\theta_6 = \tan^{-1}\left(\frac{-C_5[C_{234}(C_1o_x + S_1o_y) + S_{234}o_z] + S_5(S_1o_x - C_1o_y)}{-S_{234}(C_1o_x + S_1o_y) + C_{234}o_z}\right)$$

# Appendix C: Lagrangian Analysis of a Two-link Mechanism

The Lagrangian function $L$ may be defined as the difference between the kinetic energy $K$ and the potential energy P of the system.

$$L = K - P$$

The dynamic equations take the form (Wellstead, 1979)

$$\frac{\mathrm{d}}{\mathrm{d}t}\left(\frac{\partial L}{\partial \dot{q}_i}\right) - \frac{\partial L}{\partial q_i} + \frac{\partial C}{\partial \dot{q}_i} = F_i, \quad i = 1, n$$

where $n$ is the number of degrees of freedom, $q_i$ is the co-ordinate in which the $i$th kinetic and potential energies are expressed, $C$ is the co-content of the system dissipators, and the $F_i$ are the generalised forces. For the time being, the dissipated energy will be neglected.

As an example, figure C.1 shows a schematic of the two horizontal links of a SCARA type robot (see figures 1.4 and 2.8 for actual examples). We will assume that the masses of the links are negligible in comparison with the masses of the actuators located at the joints, and that the movements in all other joints have no dynamic effect on the two horizontal links.

With reference to figure C.1, the kinetic energy $K$ is given by

$$K = \tfrac{1}{2}m_1v_1^2 + \tfrac{1}{2}m_2v_2^2$$

Using the cosine rule, $v_2$ may be expressed in terms of $q_1$ and $q_2$, giving

$$K = \tfrac{1}{2}m_1(a_1\dot{q}_1)^2 + \tfrac{1}{2}m_2\{(a_1\dot{q}_1)^2 + (a_2\dot{q}_2)^2 + 2a_1a_3\dot{q}_1\dot{q}_2 \cos \alpha\}$$

Since the potential energy is constant and may be conveniently defined as being zero, then the Lagrangian $L = K$.

For co-ordinate $q_1$, since there are no dissipative elements, Lagrange's equation is given by

$$\frac{\mathrm{d}}{\mathrm{d}t}\left(\frac{\partial K}{\partial \dot{q}_1}\right) - \frac{\partial K}{\partial q_1} = F_1$$

and thus after further expressing $a_3$ and $\alpha$ in terms of the other variables

$$F_1 = (m_1 + m_2)a_1^2\ddot{q}_1 + m_2a_1a_2(\ddot{q}_2 \cos q_2 - \dot{q}_2^2 \sin q_2)$$
$$+ m_2a_2(a_2\ddot{q}_1 + 2a_1\ddot{q}_1\cos q_2 - 2a_1\dot{q}_1\dot{q}_2 sin\ q_2 + a_2\ddot{q}_2)$$

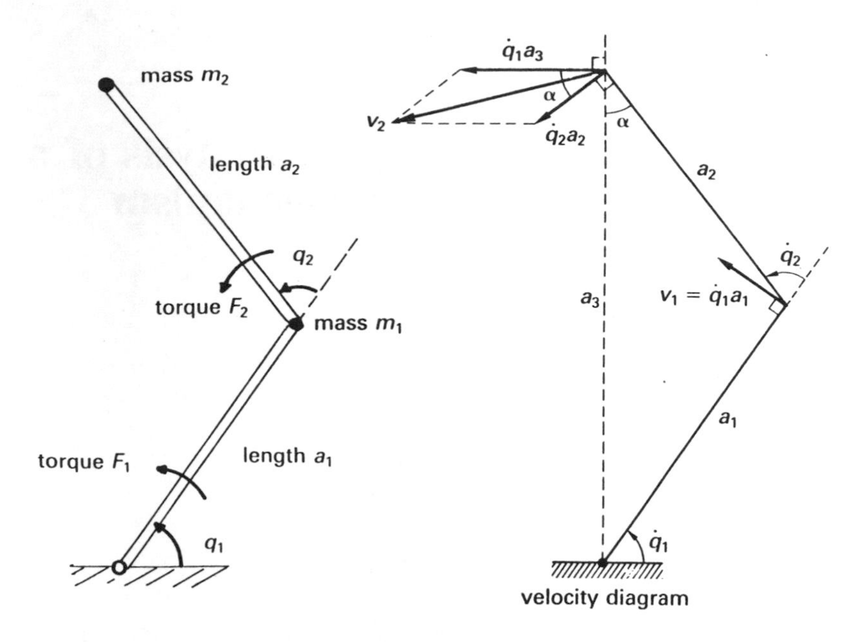


Figure C.1

For co-ordinate $q_2$, Lagrange's equation is

$$\frac{\mathrm{d}}{\mathrm{d}t}\left(\frac{\partial K}{\partial \dot{q}_2}\right) - \frac{\partial K}{\partial q_2} = F_2$$

giving

$$F_2 = m_2 a_2\{a_2(\ddot{q}_1 + \ddot{q}_2) + a_1\ddot{q}_1 \cos q_2 + a_1^2\dot{q}_1^2 \sin q_2\}$$

# References

Acarnley, P. P. (1982). *Stepping Motors: a Guide to Modern Theory and Practice*, IEE Control Engineering Series No. 19, Peter Peregrinus, Stevenage, UK.

Agin, G. J. (1985). 'Vision systems', in S.Y. Nof (Ed.), *Handbook of Industrial Robotics*, Wiley, New York, pp.231–261.

Aleksander, I. and Burnett, P. (1983). *Reinventing Man*, Pelican Books, London.

Ambler, A. P., Barrow, H. G., Brown, C. M., Burstall, R. M. and Popplestone, R. J. (1975). 'A versatile system for computer controlled assembly', *Artificial Intelligence*, **6**(2), pp.129–156.

Ambler, A. P., Popplestone, R. J. and Kempf, K. G. (1982). 'An experiment in the offline programming of robots', *Proc. 12th ISIR, Paris*, IFS (Publications), Kempston, UK, pp.491–504.

Andreasen, M. M., Kahler, S., Lund, T. and Swift, K. G. (1988). *Design for Assembly*, 2nd edn, IFS (Publications), Kempston, UK.

Asimov, I. (1983). *The Complete Robot*, Panther Books, London.

Barry Wright Corp. (1984). *Sensoflex Product Data*, 700 Pleasant St, Watertown, Massachusetts, USA.

Batchelor, B. G., Hill, D. A. and Hodgson, D. C. (1985). *Automated Visual Inspection*, IFS (Publications), Kempston, UK.

Bavarian, B. (1988). 'Introduction to neural networks for intelligent control', *IEEE Control Systems Magazine*, **8**(2) April, pp.3–7.

Baylou, P. (1987). 'Agricultural robots', *Proc. IFAC 10th World Congress, Munich*, Vol. 5, pp.114–122.

Bayo, E. (1988). 'Computed torque for the position control of open-chain flexible robots', *Proc. IEEE Conference on Robotics and Automation, Philadelphia*, pp.316–321.

Bellman, R. and Kalaba, R. (Eds) (1964). *Mathematical Trends in Control Theory*, Dover, New York.

Benett, S. (1979). *A History of Control Engineering*, Peter Peregrinus, Stevenage, UK.

Bergman, S., Jonsson, S., Ostlund, L. and Brogardh, T. (1986). 'Real-time, model-based parameter scheduling of a robot servo-control system', *Proc. 16th ISIR, Brussels*, IFS (Publications), Kempston, UK, pp.321–334.

Bonner, S. and Shin, K. G. (1982). 'A comparative study of robot languages', *Computer*, Dec., IEEE, New York, pp.82–86.

Braggins, D. and Hollingum, J. (1986). *The Machine Vision Sourcebook*, IFS (Publications), Kempston, UK.

Burgess, D. C., Hill, J. J. and Pugh, A. (1982). 'Vision processing for robot inspection and assembly', *SPIE Proc. Robots and Industrial Inspection*, Vol. 360.

Choi, B. S. and Song, S. M. (1988). 'Fully automated obstacle-crossing gaits for walking machines', *Proc. IEEE Conference on Robotics and Automation, Philadelphia*, pp.802–807.

Coppock, G. E. (1984). 'Robotic principles in the selective harvest of Valencia oranges', *Robotics and Intelligent Machines in Agriculture*, ASAE, pp.138–145.

Cronshaw, A. J., Heginbotham, W. B. and Pugh, A. (1980). 'A practical vision system for use with bowl feeders', *Proc. 1st Conf. on Assembly Automation, Brighton, UK.*, IFS (Publications), Kempston, UK, pp.265–274.

Denavit, J. and Hartenberg, R. S. (1955). 'A kinematic rotation for lower-pair mechanisms based on matrices', *ASME Journal of Applied Mechanics*, June, pp.215–221.

Dickinson, H. V. (1963). *A Short History of the Steam Engine*, 2nd edn, Frank Cass, London.

Duff, M. J. (1982). 'CLIP 4', in K.S. Fu and T. Ichikawa (Eds), *Special Computer Architecture for Pattern Recognition*, CRC Press, Boca Rotan, Florida, USA.

Durrant-Whyte, H. F. (1988). *Integration, Coordination and Control of Multi-Sensor Robot Systems*, Kluwer Academic Press.

Edwall, C. W., Ho, C. Y. and Pottinger, H. J. (1982). 'Trajectory generation and control of a robot arm using spline functions', *Proc. Robots 6*, Society of Manufacturing Engineers, Dearborn, Michigan, USA.

Electrocraft Corp. (1980). *DC Motor Speed Controls Servo Systems*, 5th edn, 1600 Second St South, Hopkins, Minnesota, USA.

Fenton, R. G., Benahabib, B. and Goldenberg, A. (1983). 'Performance evaluation of seven degrees of freedom robots, *Proc. 2nd IASTED Symp. on Robotics and Automation, Lugano*, ACTA Press, Calgary, pp.152–155.

Fischetti, M. A. (1985). 'Robots do the dirty work', *IEEE Spectrum*, April, pp.65–72.

Franklin, G. F., Powell, J. D. and Emami-Naeini, A. (1986). *Feedback Control of Dynamic Systems*, Addison-Wesley, Reading, Massachusetts, USA.

Fresonke, D. A., Hernandez, E. and Tesar, D. (1988). 'Defection prediction for serial manipulators', *Proc. IEEE Conf. on Robotics and Automation, Philadelphia*, pp.482–487.

General Motors (1985). *MAP Specification*, Version 2.1.

Grand d'Enson, A. (1985). 'Robotic harvesting of apples', *Agrimation I*, ASAE, pp.210–214.

Harmon, L. D. (1982). 'Automated tactile sensing', *Int. J. Robotics Res.*, **1**(2), pp.46–61.

Hewit, J. R. (1983). 'The robot control problem', in A. Pugh (Ed.), *Robotic Technology*, Peter Peregrinus, London, pp.15–28.

Horner, G. R. and Lacey, R. J. (1982). 'High performance brushless PM motors for robotics and actuator applications', *Proc. 1st European Conf. on Electrical Drives/Motors/Controls*, IEE Conference Publications No. 19.

Houpis, C. H. and Lamont, G. B. (1985). *Digital Control Systems, Theory, Hardware, Software*, McGraw-Hill, New York.

Inigo, R. M., Minnix, J. I., Hsin, C. and McVey, E. S. (1986). 'A biological-structure visual sensor for robotics applications', *Proc. 16th ISIR, Brussels*, IFS (Publications), Kempston, UK, pp.491–502.

Inmos Ltd (1986). *Transputer Reference Manual*, Inmos, Bristol, UK.

Jarvis, R. A. (1983). 'A perspective on range finding techniques for computer vision', *IEEE Trans. Pattern Analysis and Machine Intelligence*, **PAMI-5**, pp.122–139.

Jones, J. D. C. and Jackson, D. A. (1986). 'Physical measurement using monomode optical fibres', *Proc. IFAC Symp. on Low Cost Automation, Valencia*, pp.55–60.

Kelley, R. B., Martins, H. A. S., Birk, J. R. and Dessimoz, J. D. (1983). 'Three vision algorithms for acquiring work pieces from bins', *Proc. IEEE*, **21**(7), pp.803–820.

Kemp, D. R., Taylor, G. E., Taylor, P. M. and Pugh, A. (1986). 'A sensory gripper for handling textiles', in D. T. Pham and W. B. Heginbotham (Eds), *Robot Grippers*, IFS (Publications), Kempston, UK, pp.155–164.

Krutz, L. J., Miles, G. E. and Harmmer, P. A. (1986). 'Robotic transplantation of bedding plants', *Agrimation II*, ASAE, pp.78–80.

Loughlin, C., Taylor, G. E., Pugh, A. and Taylor, P. M. (1980). 'Visual inspection package for automated machinery', *Proc. Automated Inspection and Product Control, Stuttgart*, IFS (Publications), Kempston, UK, pp77–85.

Luh, J. Y. S. (1983). 'Conventional controller design for industrial robots — a tutorial', *IEEE Trans. on Systems, Man, and Cybernetics*, **SMC 13**(3), pp.298–316.

MacFarlane, A. G. J. (1979). *Frequency Response Methods in Control Systems*, IEEE Press, New York.

Makino, H. and Furuya, N. (1980). 'Selective compliance assembly robot arm', *Proc. 1st Int. Conf. on Assembly Automation, Brighton*, IFS (Publications), Kempston, UK, pp.77–86.

Margrain, P. (1983). 'Servo-actuators and robots', in *Developments in Robotics*, IFS (Publications), Kempston, UK.

Maxwell, T. C. (1868). 'On governors', *Proc. Royal Society*, **16**, pp.270–283.

Mayr, O. (1970). *The Origins of Feedback Control*, MIT Press, Cambridge, Massachusetts, USA.

McCloy, D. and Harris, M. (1986). *Robotics, an Introduction*, Open University Press, Milton Keynes, UK.

McCloy, D. and Martin, H. R. (1980). *The Control of Fluid Power*, Ellis Horwood, Chichester, UK.

Microbot Inc. (1982). *Operation of the 5 Axis Table-top Manipulator MIM-5*, Microbot, Mountain View, California, USA.

Micro-Robotics Ltd (1984). *The Snap Camera Manual*, Micro-Robotics, Cambridge, UK.

Monsion, M. *et al.* (1985). 'Visual servo-control in automatic winter pruning of the grapevine', *Proc. IEE Conf. Control 85, Cambridge*, pp.399–404.

Montalescot, J. B. (1986). 'L'electronique et les automatismes en élevage laitier: vers la robotisation de la traite', *Agrimation II*, ASAE.

Morgan, A. T. (1979). *General Theory of Electrical Machines*, Heydon, London.

Mott, D. H., Lee, M. H. and Nicholls, H. R. (1984). 'An experimental very high resolution tactile sensor array', *Proc. RoViSeC 4*, IFS (Publications), Kempston, UK, pp.241–250.

Nyquist, H. (1932). 'Regeneration theory', *Bell Sys. Tech. J.*, **11**, pp.126–147.

Paul, R. P. (1981). *Robot Manipulators, Mathematics, Programming and Control*, MIT Press, Cambridge, Massachusetts, USA.

Peyran, B. *et al.* (1986). 'Description d'un robot autonome destiné à la taille automatique de la vigne', *Agrotique 86*, ADESO.

Pham, D. T. and Heginbotham, W. B. (Eds) (1986). *Robot Grippers*, IFS (Publications), Kempston, UK.

Pugh, A., Heginbotham, W. B. and Page, C. J. (1977). 'Novel techniques for tactile sensing in a three dimensional world', *The Industrial Robot*, March, pp.18–26.

Raibert, M. H. (1986). *Legged Robots that Balance*, MIT Press, Cambridge, Massachusetts, USA.

Ránky, P. G. and Ho, C. Y. (1985). *Robot Modelling — Control and Applications with Software*, IFS (Publications), Kempston, UK.

Redford, A. and Lo, E. (1986). *Robots in Assembly*, Open University Press, Milton Keynes, UK.

Rosenbrock, H. H. (1979). 'The reduction of technology', *Proc. IFAC Symp. on Criteria for selecting appropriate technologies under different cultural, technical and social conditions, Bari, Italy*, pp.1.2–1 to 1.2–7.

Stewart, D. (1965). 'A platform with six degrees of freedom', *Proc. Inst. Mech. Eng.,* **180**(Pt 1)(15), pp.371–386.

Stringer, J. D. (1976). *Hydraulic Systems Analysis*, Macmillan, London.

Taylor, G. E., Kemp, D. R., Taylor, P. M. and Pugh, A. (1982). 'Vision applied to the orientation of embroidered motifs in the textile industry', *Proc. RoViSeC 2, Stuttgart.*

Taylor, G. E., Taylor, P. M., Gibson, I. and Shen, S. (1987). 'Optimal use of low resolution vision sensors', *Proc. 6th IASTED Symp. on Modelling, Identification and Control, Grindelwald.*

Taylor, P. M. (1988). 'Multisensory assembly and error recovery', in G. E. Taylor (Ed.), *Kinematic and Dynamic Issues in Sensor Based Control*, NATO ASI Series, Springer-Verlag, Berlin.

Taylor, P. M. and Bowden, P. (1986). 'The use of multiple low cost vision sensors in fabric pick and place tasks', *Proc. IFAC Symp. on Low Cost Automation, Valencia*, pp.89–95.

Taylor, P. M. and Gilbert, J. (1987). 'Nonlinear control for robots with significant drive nonlinearities', *Proc. IEEE Conf. on Decision and Control, Los Angeles*, pp.1035–1036.

Taylor, P. M. and Koudis, S. G. (1987). 'Automated handling of fabrics', *Science Progress*, **71**, pp.353–365.

Taylor, R. H., Summers, P. D. and Meyer, J. M. (1983). 'AML a manufacturing language', *Int. J. Robotics Res.*, **1**(3), pp.19–41.

Thornton, G. S. (1988). 'The GEC Tetrabot — a new serial–parallel assembly robot', *Proc. IEEE Conference on Robotics and Automation, Philadelphia*, pp.437–439.

Truckenbrodt, A. (1981). 'Truncation problems in the dynamics and control of flexible mechanical systems', *Proc. IFAC 8th World Congress, Kyoto*, Vol. 14, pp.60–65.

Unimation Inc. (1985). *Unimate Industrial Robot Programming Manual. Users' Guide to VAL II Vers. 1.4B*, Unimation, Danbury, Connecticut, USA.

Usher, M. J. (1985). *Sensors and Transducers*, Macmillan, London.

Waize, H. (1986). 'Low-cost LAN supplements MAP for real time applications', *Proc. IFAC Symp. on Low Cost Automation, Valencia*, pp.171–176.

Wampler, C. W., II. (1989). 'The inverse function approach to sensor-driven kinematic control of redundant manipulators', in G. E. Taylor (Ed.), *Kinematic and Dynamic Issues in Sensor Based Control*, Springer-Verlag, Berlin.

Wellstead, P. E. (1979). *Introduction to Physical System Modelling*, Academic Press, London.

Whitney, D. E. (1986). 'Part mating in assembly', in S. Y. Nof (Ed.), *Handbook of Industrial Robots*, Wiley, New York, pp.1086–1116.

Whitney, D. E. and Nevins, J. L. (1979). 'What is the remote center compliance and what can it do?', *Proc. 9th ISIR, Washington DC*, pp.135–152.

Williams, S. J. (1985). 'Multivariable controller design for industrial robots', *Proc. IEE Conference Control 85*, IEE Conference Publication 252, pp.309–314.

Wong, P. C. and Hudson, P. R. V. (1983). 'The Australian robotic sheep shearing research and development program', *Proc. Robots 7*, pp.10–56 to 10–63.

Yasutoru, F., Takuoka, D., Yamada, M. and Tsukamoto, K. (1988). 'Cleaning robot control', *Proc. IEEE Conference on Robotics and Automation, Philadelphia*, pp.1839–1841.

Yin, B., Ambler, A. P. and Popplestone, R. J. (1984). 'The use of vision verification in a high level offline programming language', *Proc. RoViSeC 4, London*, IFS (Publications), Kempston, UK, pp.371–377.

Yong, Y. F., Gleave, J. A., Green, J. L. and Bonney, M. C. (1985). 'Off-line programming of robots', in S.Y. Nof (Ed.), *Handbook of Industrial Robots*, Wiley, New York, pp. 366–380.

Young, K. K. D. (1978). 'Controller design for a manipulator using theory of variable structure systems', *IEEE Trans. Systems, Man, and Cybernetics*, **SMC-8**(2), pp.101–109.

Zarrop, M. B. (1987). 'Generalized pole-placement self-tuning control of a robot manipulator', *IEE Colloquium on Recent Advances in Robot Control*, IEE Digest No. 1987/22, pp.6/1 to 6/7.

# Index